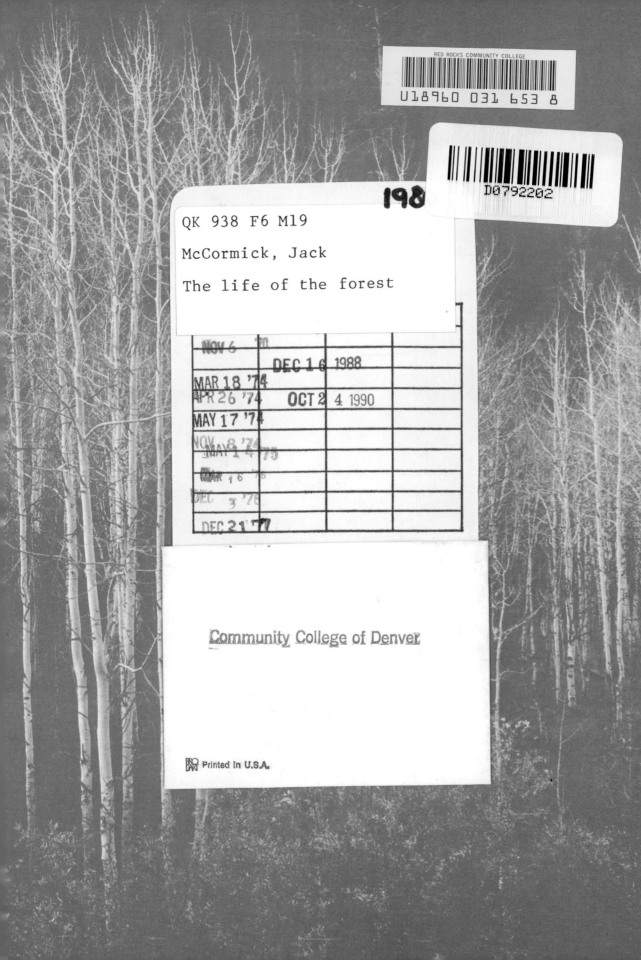

OUR LIVING WORLD OF NATURE

The
Life
of the
Forest

Developed jointly with The World Book Encyclopedia

Produced with the cooperation of
The United States Department of the Interior

OUR LIVING WORLD OF NATURE

The
Life
of the
Forest

JACK McCORMICK

Published in cooperation with
The World Book Encyclopedia

McGraw-Hill Book Company
NEW YORK   TORONTO   LONDON

JACK McCORMICK *is Curator and Chairman of the Department of Ecology and Land Management at the Academy of Natural Sciences of Philadelphia and Resident Director of the Academy's Waterloo Mills Field Research Station at Devon, Pennsylvania. He is also lecturer in the Department of Biology and Department of Landscape Architecture and Regional Planning at the University of Pennsylvania. Educated at Butler University in Indiana and at Rutgers, The State University of New Jersey, where he was awarded a doctorate in 1955, the author was a member of the scientific staff of the American Museum of Natural History in New York City from 1954 to 1961. His first book on forest ecology,* The Living Forest *(Harper & Row, 1959), was based on information gathered for the Museum's Hall of North American Forests, which he supervised. Now Consultant in Ecology, he directs vegetation studies at the Museum's Kalbfleisch Field Research Station on Long Island. From 1961 to 1963, Dr. McCormick was Assistant Professor of Botany and Research Associate in Polar Studies at Ohio State University. His primary interest is the forests of North America, and during the past decade his work has taken him to forty-two states, Mexico, Canada, and Greenland.*

Library of Congress Catalog Card Number: 66-14241

7890 NR 721069
46001

OUR LIVING WORLD OF NATURE

Science Editor

    RICHARD B. FISCHER   *Cornell University*

Board of Consultants

ROLAND CLEMENT	*National Audubon Society*
C. GORDON FREDINE	*National Park Service, The United States Department of the Interior*
WILLIAM H. NAULT	*Field Enterprises Educational Corporation*
BENJAMIN NICHOLS	*Cornell University*
EUGENE P. ODUM	*University of Georgia*
HENRY J. OOSTING	*Duke University*
OLIN SEWALL PETTINGILL, JR.	*Cornell University*
DAVID PIMENTEL	*Cornell University*
PAUL B. SEARS	*Yale University*
ROBERT L. USINGER	*University of California*

Readability Consultant

    JOSEPHINE PIEKARZ IVES   *New York University*

Special Consultants

ALEXANDER B. KLOTS	*City College of New York*
ROBERT T. ORR	*California Academy of Sciences*
CHARLES E. ROTH	*Massachusetts Audubon Society*
EARL L. STONE	*Cornell University*

Contents

LAND OF MANY FORESTS

APPENDIX

The Forest
Community

Stop for a minute in the shade of this great tree. Is it an oak?
If you are in California it may be a giant redwood. If you
are in the Great Smoky Mountains it may be a tulip tree,
rising 150 feet, straight as a marble column. Elsewhere it
may be an ash, a sycamore, a great fir, or a maple—any of
the many splendid trees that grow in American forests. For
the moment at least, the name doesn't really matter.

Sit down, lean back. Look at the forest around you. The
light is dim, but here and there slanting shafts of sunlight
fall through the canopy of leaves like light through a cathedral window. A royal-purple butterfly flits among the dusky
tree trunks. Vines hang in long, graceful loops; scattered
below them are the green leaves of saplings and shrubs. On
the ground you see a carpet of dead leaves, interrupted perhaps by occasional patches of wintergreen and partridgeberry, colonies of meandering ground pine, or cushions of
velvety moss. A clump of ferns catches the light from a sun-
ray and glows for a moment like a fountain of fire.

At first the forest seems quiet. Compared to a city street,
it *is* quiet, yet soon you begin to hear the forest's many
voices. High overhead, leaves murmur in the wind. Insects

9

The eight-eyed wolf spider lives in forest clearings, under stones or in holes. It does not spin webs, but chases down insects on the ground. If frightened, it runs rapidly on its hairy legs and may even go underwater to escape an enemy.

Buried up to its bulbous eyes in a clump of moss, a Pacific tree frog lies in ambush for an insect meal. It will snatch its prey out of the air with a lightning-quick flick of its sticky tongue.

buzz in the air. A squirrel chatters. Startlingly close, an ovenbird utters its vibrant cry, and in the distance a dove coos mournfully. Something creates a disturbance among the dead leaves not far away and makes enough noise to be a bear; probably it is only a hermit thrush or a ruffed grouse. A deer, twitching its ears and flicking its white tail, stops at the edge of a glade to look at you.

You are becoming aware of the life around you. The tree you lean against is as alive as you are. Gradually you realize that the forest is a community composed of millions of living things—in the ground, in the air, among the leaves and branches. How many hidden eyes are watching you? Perhaps hundreds, though by far the largest number of the forest's inhabitants care nothing about your intrusion. On

10

a pleasant summer afternoon it doesn't seem to matter how many wild creatures are surveying you. Indeed, you feel quite at home in this vast community, one creature among the great fraternity of the living.

Only a few hundred years ago, the land surface of much of the world was covered with forest. The communities of men were surrounded and separated by tree-covered wilderness. Men's thoughts and feelings were constantly shaped by the image of the deep woods. In mythology the forest was the domain of elves, witches, dragons, giants, and demons; it also became the place of adventure and romance—the home of Robin Hood, the hunting ground of Hiawatha.

Today in the United States the boundless primeval forest is gone. Only in specially preserved areas such as the great

Not eyes at all but merely markings, the conspicuous oval spots on the body of the spicebush swallowtail caterpillar may help to frighten off hungry birds.

The marten, a member of the weasel tribe, is a sharp-eyed hunter agile enough to chase down red squirrels, sometimes to the very tips of evergreen branches. Equally at home in trees or on the forest floor, martens eat not only squirrels and other small animals, but also fruits, nuts, and berries.

national parks can you find remnants of the original wilderness. But the forest is still with us. Indeed, about a third of the United States is covered with trees. Wherever there is a wood lot or an untended hillside, you will find the living community of the forest in one stage or another of its development.

The trees

Many plants grow in the forest, of course, but the most conspicuous are the trees. For this reason we name forests after the trees that most commonly grow in them. We often speak of a pine woods or a birch grove. In New England many farmers have *sugarbushes,* forests in which sugar maples are the most abundant trees. Scientists use the same system: they refer to forests by the name of the predominant trees. Thus it is common to speak of beech–maple forests, spruce–fir forests, or oak–hickory forests.

If you want to observe the forest community intelligently, you must know something about trees. Not necessarily everything; you don't need to become an expert *dendrologist* (a dendrologist is a botanist who specializes in the study of trees). But you should recognize the main groups and common species.

The scientific distinctions are very precise, although scientists sometimes disagree among themselves about details. Ask them the basic question *What is a tree?* and you will get varying answers. But for our purpose we can accept the common, practical definition. A tree is a woody plant twelve feet or more tall with a single main stem, or trunk, and a more or less distinct crown of leaves. Woody plants that are smaller or many-stemmed are called shrubs.

Trees are divided into two main groups, *coniferous* trees and *flowering* trees. Most coniferous trees, such as pines and hemlocks, bear their seeds in cones. Juniper berries are actu-

Thirty-one national parks, covering more than 13,500,000 acres, preserve splendid examples of original forestland. Isle Royale National Park *(top)* occupies the largest island in Lake Superior and is blanketed with mixed forests of coniferous and deciduous trees. Mount Rainier National Park *(bottom),* covering 377 square miles in western Washington, contains dense forests of spruces and hemlocks nestled between rugged mountain peaks.

On the lookout, a red fox eyes the forest floor for an unwary mouse, squirrel, or other small animal. In the summer and autumn it varies its meaty fare with forest fruits and berries. Red foxes may be red, tan, or black; and the fur of darker individuals, like the one pictured here, may be tipped with silver.

DECIDUOUS

CONIFEROUS

ally modified cones. The scarlet berries of yew, however, are quite different structures. Most conifers have needlelike leaves and are evergreen, although the needles of a few, such as the tamarack and the bald cypress, drop off each autumn. Most of North America's native flowering trees are *deciduous*—that is, all their leaves fall in autumn—and all have broad leaves, like oaks and maples, rather than needle-like leaves. We commonly refer to them as deciduous or *broadleaf* trees. Some, such as hollies and live oaks, however, are evergreen rather than deciduous.

For our purposes, this general division into coniferous (needleleaf) and broadleaf trees will be sufficient most of the time. It also coincides with the distinction made by lumbermen and foresters between softwoods and hardwoods. Most coniferous trees are softwoods, and lumber cut from them is used for rough construction such as the framework of houses. Broadleaf trees make hard lumber, used for furniture, tool handles, and high-grade flooring. Even so, this distinction is far from exact, since some so-called softwoods, such as Douglas fir, are actually harder than some so-called hardwoods, such as basswood.

Actually, these distinctions would have to be much more detailed to make them rigidly correct. Evergreens do shed their leaves, for instance, even the coniferous trees; they just do it more slowly than other trees, and not all at one time. In tropical forests, the broadleaf trees may stay green all year. Yet for the temperate forests of North America, with which we will be dealing in this book, our categories fit well enough. *Broadleaf* and *coniferous*, *deciduous* and *evergreen*, *hardwood* and *softwood*—these are the terms actually used by scientists and by rangers and lumbermen who live and work in the forest.

In a deciduous forest *(above)*, there is a tremendous variety of vegetation. Broadleaf beeches and maples tower over tree seedlings, shrubs, wild flowers, and a forest floor cluttered with last year's leaves. In a coniferous forest *(below)*, there is less variety; few leafy herbs and shrubs grow on the forest floor. Like pillars in a cathedral, conifers tower over a carpet of needles interspersed with occasional seedlings that have sprung from the seeds of the predominant parent trees.

The food machines

The trees of the forest may be merely so much standing lumber or they may be mysterious sources of beauty and inspiration, depending upon the interests of the observer. But they are always one thing more: they are food machines. During the warm days of summer, the trees—and all the other green plants of the forest—are engaged in making the basic food of life.

This food is a simple form of sugar known as glucose. The process by which plants make this sugar is called photosynthesis. *Photosynthesis* is a word that comes from two Greek words meaning "put together with light"; it is the process that links us and nearly all living organisms with the earth's ultimate source of energy, the sun. Indeed, were it not for this remarkable process, life as we know it could not exist.

Many things about photosynthesis are still understood only dimly by scientists, but we do know that green plants use water from the soil and carbon dioxide from the air, and in the presence of sunlight change these substances into sugar. By means of further chemical changes, all of which take place in plants and some of which take place in animals, this basic sugar is converted into more complex carbohydrates, proteins, fats, and other substances that support life.

Photosynthesis involves a number of extremely complicated chemical reactions that occur in the tiny bodies containing the green coloring matter of plants, *chlorophyll*. But you need only remember that photosynthesis is the process by which green plants convert the energy of sunlight into chemical energy in sugar and other compounds and thus produce food for themselves and for nearly all other living things. Of course, not every animal eats plants, yet the owl that eats a shrew that ate a caterpillar that ate a leaf depends upon plants just as much as any plant-eating animal does.

A CHLOROPLAST ENLARGED 6000 TIMES

Photosynthesis—the vital process by which plants produce food for themselves and for nearly all other living things—takes place within plant cells in tiny bodies called chloroplasts. These minute food factories contain sandwiched layers of chlorophyll, the green coloring matter of plants. When sunlight strikes a chloroplast, a chain of chemical reactions occurs almost instantaneously. Water from the soil and carbon dioxide from the air are converted into glucose, a simple sugar. This basic sugar is subsequently used to build more complicated food substances. Each unit of food contains a small quantity of solar energy stored as chemical energy. Later, when a plant or animal "burns" the food in respiration, much of this energy is released as heat.

Only about 1 per cent of the sunlight falling on these sugar-maple leaves is actually used in the food-making process of photosynthesis. Nevertheless, if the light energy captured by one acre of beech–maple forest in a year were converted into electrical energy, it would equal the power consumed annually by some fifty average American homes.

Leaf-cutting ants carry leaf fragments to their underground nests and thus provide food for a fungus which produces "buds" the ants eat.

The pine panthea moth caterpillar feeds on white-pine needles by bending them into loops between its front and rear legs and eating them from the tips inward.

One by-product of photosynthesis is oxygen. As glucose is formed from water and carbon dioxide, oxygen from the water molecules is released into the air. Since nearly all plants and animals, including man, need oxygen to live, and since no animal can release oxygen, the supply soon would be exhausted if the plants did not continually replenish it. Green plants are, therefore, the foundation upon which the rest of life is built, for they are the source of all the food we eat and they release the oxygen we breathe.

The eaters and the eaten

Animals flourish wherever they find an abundance of food. As a result, the forest is a rich environment for many animals. Deer browse on herbs, shrubs, and the lower branches of trees; mice and squirrels eat seeds, buds, and bark; caterpillars chew up great quantities of foliage every day.

No part of a plant is safe from animal attack. Leaves, seeds, and roots are commonly preferred by the plant-eaters, but even the trunks of the hardest trees can provide nourishment for some of them. Many animals feed only on a particular part; some cicada larvae, for instance, suck the juices from roots. Others, such as chipmunks and squirrels, may eat several parts of a plant.

Similarly, some animals feed on many kinds of plants,

High in the branches of a spruce, a red squirrel inspects an opened cone for seeds. Earlier in the year, this busy rodent dropped hundreds of green cones to the ground and buried them in mounds, some containing as many as eight to ten bushels. Eventually it will dig up most of the cones, cut off the scales, and eat the seeds. It will not retrieve every cone, however, and thus the red squirrel helps to distribute and plant spruce seeds.

while the diets of others are limited to a single kind. White-pine weevils usually bore into only the tender new shoot (called the *leader*) that grows in spring at the top of the main stem of a white pine. Thus the white-pine weevil is a highly specialized type of plant-eater, able to live only where white pines and a few other conifers grow.

The white-tailed deer, on the other hand, browses first on one kind of tree, then on another, as well as on leafy weeds, grasses, and ferns. Deer will eat not only leaves but also acorns, fruits, twigs, beechnuts, bark—almost anything. Nevertheless, plant-eaters tend to be specialists, like the weevils, rather than nonspecialists, like the deer.

The plant-eaters themselves are in turn a source of food, for in the forest as elsewhere, there are many types of animals, called *predators*, that live by capturing and eating the plant-eaters. In turn, there are other, usually bigger predators that live on the predators that live on plant-eaters. Cankerworms, for instance, feed on the leaves of trees. Red-eyed vireos eat cankerworms. Sharp-shinned hawks eat vireos. This is a *food chain* in which the first link is the plant, the second link is the plant-eater (in this case the canker-worm), and the third and fourth links are the predators (vireos and hawks).

But because plants and animals must continually use chemical energy to stay alive, there is less energy available

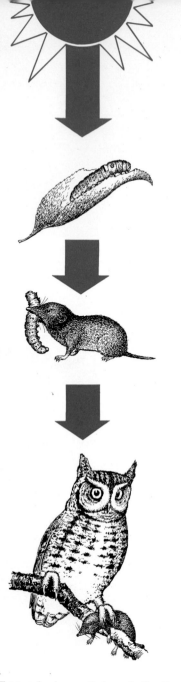

Largest of all the North American cats, the cougar, or mountain lion, hunts deer and other mammals. Moving with soundless stealth, it creeps to within a few feet of its prey. Then it springs, hurling its full weight (which may be more than 200 pounds) against its hapless victim, sinks its teeth into the animal's neck, and drags it to the ground.

Every animal on earth depends directly or indirectly upon green plants for its food. Here a wild red-cherry leaf, which has converted oxygen and water into food, is being eaten by a tiger swallowtail caterpillar. Later the caterpillar falls victim to a predator, a short-tailed shrew. Finally, the shrew is eaten by another, larger predator, a screech owl. This sequence is only one of many thousands of interlocking forest food chains.

as we go up the chain. When cankerworms eat leaves, they get less energy than the leaves originally captured from the sun. When red-eyed vireos eat cankerworms, they get less energy than the cankerworms got from the leaves. When hawks eat red-eyed vireos, they get less energy than the vireos got from the cankerworms. All along the chain chemical energy is constantly being used. Most is changed to heat energy and is lost to the air.

Just as the amount of usable energy decreases, so does the number of animals at each food level. In the forest there are more plants (or at least a greater mass of plant material) than plant-eaters, more plant-eaters than predators, and more small predators than big ones. No animal population can increase faster than its source of food for long. Because it takes more food to feed a hawk than to feed a vireo, and because less food per acre is available to the hawk, the number of hawks in any area is likely to be smaller than the number of vireos or other insect-eating birds.

Also, as we move up the food chain there is a progression from specialists to nonspecialists. Among the plant-eaters, particularly the insects, most are specialists, living on one particular kind of plant. But among the predators the diets of comparatively few are limited to one kind of animal. Sharp-shinned hawks prey upon many songbirds besides vireos, and upon mice, rats, squirrels, snakes, and dozens of other animals as well.

One type of predator that usually is specialized has solved the problem of getting food very well. It lives on, or in, its prey, so that it has a readily available food supply. These predators are called *parasites*, and there are thousands of different kinds—among them fleas, ticks, worms, mites, fungi, bacteria. Every animal has parasites living in its fur, inside its stomach or intestines, in its blood, even in its muscle tissue. Usually the parasites are far smaller than the animals they attack, but they are also far more numerous.

20

Thrusting upward through the litter, a yellow coral mushroom stands out boldly against the muted colors of the forest floor. Unlike green plants, which produce their own food, mushrooms and other fungi, most bacteria, and a few flowering plants gain their nourishment by breaking down dead matter. In doing so, they return nitrogen, necessary for green plant growth, to the soil.

When a tree dies and falls to the forest floor **(top)**, it is attacked by a corps of forest janitors. First come the shallow-boring grubs, followed by deep borers, such as sawyer beetle larvae, powderpost beetles, and termites **(middle)**. Water seeping through tunnels these insects make speeds decomposition by fungi. Eventually the log turns into a paste in which bacteria, wireworms, and fungus beetles thrive. Finally, it is attacked by such creatures as millipedes, earthworms, and pillbugs **(bottom)**, which move in from the soil to complete the clean-up job.

Still other animals, called *scavengers*, obtain their energy from the remains and wastes of plants and animals. Several kinds of snails feed on dead plants. Vultures feed almost entirely on dead animals; crows and opossums depend on dead animals for part of their diet. Carrion beetles and sexton beetles bury the carcasses of dead animals and then lay their eggs on them. The larvae, when they hatch, feed on the dead animal tissue. Flies of many kinds also lay eggs on feces and dead animals.

More important, there are hosts of other organisms, such as millipedes, earthworms, mites, springtails, fungi, and microscopic bacteria, that live on dead material and cause its mechanical and chemical breakdown, or *decomposition.*

The work of these "decomposers" may seem insignificant, but without it life in the forest—and, for that matter, life in the world—would be impossible. The reason is simple. All living things, plants and animals, are constructed from certain basic chemical elements, such as carbon, hydrogen, oxygen, nitrogen, iron, and calcium. There is only a limited supply of these elements in the world. If the dead bodies of plants and animals, together with the wastes given off by living plants and animals, accumulated indefinitely wherever they happened to fall, large quantities of essential nutrients and energy would be tied up in an unusable form—withdrawn from circulation, so to speak. These wastes do not accumulate, however; they are ground up, chewed, dissolved, digested, and thus broken down into reusable forms again by billions of decomposers.

Every organism—every tree, shrub, insect, bird, and mammal—in the forest today is thus made from elements that once were parts of other living things. Indeed, your own body contains "secondhand materials" that probably have been used many times before and will be used many times again. Without doubt, some of the atoms now in your body were once those of a giant dinosaur that roamed the prehistoric swamps.

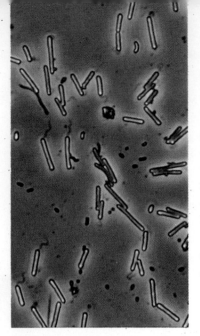

Soil bacteria such as the rod-shaped *bacilli*, here enlarged 1500 times, may number over thirty billion per pound of forest litter. Bacteria convert the remains of dead plants and animals into basic materials that can be used again by green plants.

The forest ecosystem

The forest, then, is a complex community of living things surrounded by and related intimately to nonliving things. Its living parts are plants and animals. Its ultimate source of energy is the sun. Its nonliving parts are air and water. The forest soil is both living and nonliving, since it consists of animals and plants (some of them microscopic) as well as rock particles and the dead tissues of plants and animals. All parts of the forest are intricately related in a single system which *ecologists* (scientists who study the relationship between living things and their environments) call an ecological system, or ecosystem.

This word *ecosystem* was first used in 1935 by a British scientist, Sir Arthur Tansley. He emphasized the relationships between living and nonliving parts of the community, as well as the relationships between the living parts alone. Many earlier biologists concentrated on cataloguing the living parts of the ecosystem and gave little thought to the energy and materials that pulse through it. Such specialized knowledge had resulted in great scientific advances, but sometimes it also resulted in a rather narrow view of the natural world. Sir Arthur insisted on the importance of a broader view that would include all the interworkings of a whole ecosystem.

Today most naturalists—professionals and amateurs—are committed to the ecosystem concept. It may seem confusing at first—after all, a single ecosystem can include billions of living things and vast quantities of nonliving material as well—but the parts fit together with astonishing precision. You are bewildered when you see the works of a watch spread out on a watchmaker's table, yet they can be combined into a smoothly functioning unit.

Green plants, using the sun's energy in photosynthesis, make food from inorganic substances. Portions of these plants are then consumed by plant-eaters, which are eaten in turn by predators. When plants and animals die and fall to the forest floor, fungi and bacteria feed upon their remains and decompose them into inorganic substances. The flow of matter from soil and air through plants and animals back to soil and air is a never-ending cycle upon which all life depends.

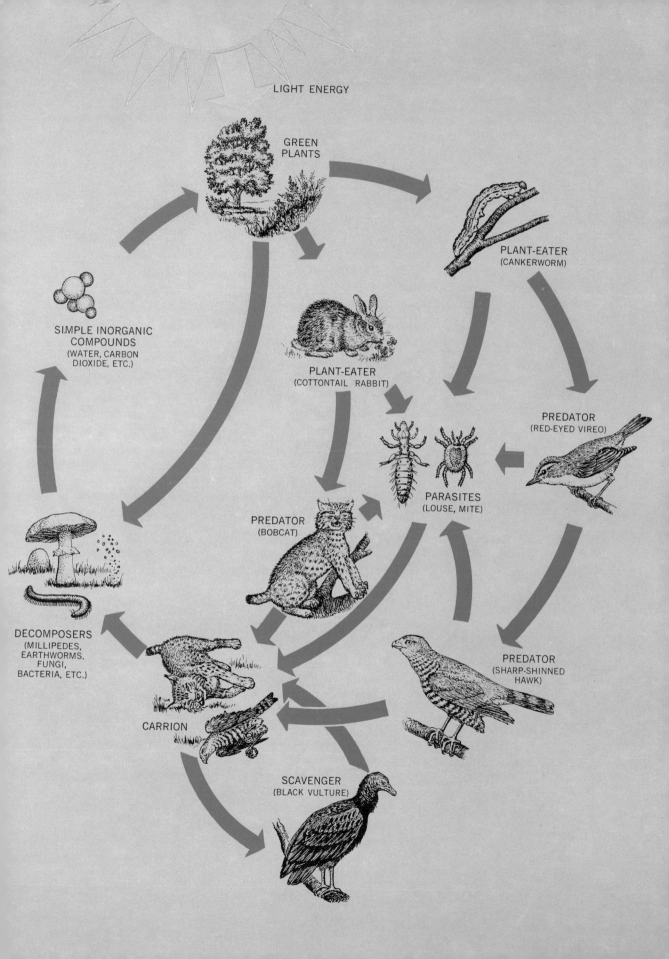

LIGHT ENERGY

GREEN PLANTS

PLANT-EATER
(CANKERWORM)

SIMPLE INORGANIC
COMPOUNDS
(WATER, CARBON
DIOXIDE, ETC.)

PLANT-EATER
(COTTONTAIL RABBIT)

PREDATOR
(RED-EYED VIREO)

PARASITES
(LOUSE, MITE)

PREDATOR
(BOBCAT)

DECOMPOSERS
(MILLIPEDES,
EARTHWORMS,
FUNGI,
BACTERIA, ETC.)

PREDATOR
(SHARP-SHINNED
HAWK)

CARRION

SCAVENGER
(BLACK VULTURE)

The microhabitat

Usually we think of the forest as a *habitat,* or place to live, for many plants and animals. We speak of forest mammals, forest insects, and forest birds. The forest, however, is not just one habitat; it is a combination of many different habitats—rotting logs, leaf litter, high branches, tree trunks, and so on.

For instance, some forest ecologists have devoted a great deal of study to tree holes. Whenever a hole is formed in the trunk of a tree, a new community springs up almost immediately. Perhaps the hole is formed by a woodpecker that uses it for a nest; or perhaps the hole begins at a place where the tree has been damaged, where the bark has been torn away and the inner tissues have begun to rot. Next year it may be enlarged by a squirrel; after that it may become the home of an owl.

The cause of the hole is not so important as what happens afterward. Very soon microscopic plants and animals begin to live in the decaying tissues of the tree, loosening them up so that a kind of "soil" forms at the bottom of the hole. Larger insects come, feeding on the microscopic animals. Then still larger types appear, until the tree hole encloses a complex little community. In general there are two broad classes of tree holes: dry and wet. If water collects in the hole, it becomes a tiny pond, and the life there will be aquatic—mosquitoes, roundworms, beetle larvae, even small frogs or toads. If the hole remains dry, the animal life is more likely to consist of such species as carpenter ants and certain types of small spiders.

In short, the tree hole is a tiny world within the forest, and scientists refer to it as a *microhabitat.* The community

Dry tree holes, often chiseled out by woodpeckers, may be used by a succession of forest animals. Following the original woodpecker, perhaps a starling, a squirrel, or a screech owl occupied the hole *(top)* before the young raccoon moved in. Wet tree holes often support teeming communities of tiny aquatic animals, including bacteria, nematode worms, insect larvae, and even tadpoles. Here, submerged in foul water, a drone fly larva *(bottom),* called a rat-tailed maggot, snorkels air through an extensible breathing tube.

26

living in a tree hole generally continues to develop until the tree dies and falls. But sometimes enough new wood and bark grow to seal a small hole, and the community is put out of operation before it gets started properly.

There are many other microhabitats in the forest, and we shall be looking at some of them in this book. A good place to begin is with the layers of the forest.

A typical mature forest may have several layers of vegetation, each supporting different kinds of animals (although some animals move around a great deal between layers). At the top, there is the *canopy*, beneath it the *understory*, then the *shrub layer*, the *herb layer*, and finally the *forest floor*.

The canopy

The leafy crowns of the forest's tallest trees make up the canopy. In some forests these tall trees are widely spaced and the canopy is open; a good deal of sunlight reaches through to the layers below. In other forests the tall trees grow close together; their branches interweave, forming a closed canopy like a huge tent that shuts out the sunlight. Whether the canopy is open or closed has much to do with what types of plants and animals flourish in the forest's lower layers.

Most of the forest's food is made in the canopy, because this is where the light is most intense and where most of the green leaves are. Photosynthesis is carried on most vigorously at the top of the forest, while the lower levels—branches, trunks, and roots—are where the food accumulates.

The upper side of the canopy is not a good habitat for most animals. There the sun's energy is too intense, the wind and rain too violent. Think of it as a world of extremes. One minute the light may be literally blinding, the next minute a passing cloud may reduce the light to a hundredth of its former intensity. During a storm the topmost branches of the canopy may be lashed by the wind so fiercely that seen from an airplane the forest resembles a storm-tossed ocean; yet at the same time the shrubs on the ground may be only gently swaying.

Just below this top surface, however, conditions are more suitable for animal life and food is plentiful. Especially in broadleaf forests, this is the zone of the leaf-eaters—thou-

UNDERSTORY

SHRUB LAYER

HERB LAYER

FOREST FLOOR

sands of different species of beetles, bugs, and caterpillars. Leaf miners are among the commonest—tiny creatures that carve tunnels between the two surfaces of a leaf and eat the soft tissue inside. Moth caterpillars, leaf hoppers, aphids, and a host of other insects make their home in the canopy.

Vireos, warblers, flycatchers, and other insect-eating birds flit continually among the canopy's branches. Spiders and dozens of kinds of predatory insects are abundant in the canopy. Porcupines sometimes inch their way into the highest branches to eat leaves and tender twigs. The climbing seed-eaters and nut-eaters, especially squirrels, are well adapted to life in the canopy.

The height of the canopy varies from forest to forest. It may be only 25 feet aboveground in the scrub-pine forests of southern New Jersey or in the piñon–juniper woodlands of the West. In a mature broadleaf forest the canopy may rise 100 to 150 feet above the floor. But a squirrel feeding on fir cones in the evergreen forests of the Pacific coast may find himself at a dizzying height of 250 feet or more.

In short, the canopy provides a feeding-ground for many

The red-banded leaf hopper often feeds among the highest tree branches. It sucks the sap from leaves and exudes from the tip of its abdomen drops of "honey-dew," which attract ants and honeybees.

Although her mate may be seen scrambling about on the forest floor, the female red tree mouse spends her life high in the branches of fir trees and seldom, if ever, descends to the ground.

The scarlet tanager feeds so high in the leafy treetops of eastern forests that it is often detected only by its buzzing *tip-churr* call. Its nest, usually built far out on a limb, may be as much as fifty feet aboveground.

creatures and at the same time furnishes protection for the forest below. Some insects spend their entire life cycles in the canopy, and a few of the most colorful songbirds, among them the scarlet tanagers and cerulean warblers, may build their nests on the high branches, where you may glimpse the flash of their plumage on a summer day. The canopy is not so dense that air and rain cannot penetrate it, but it is thick enough to reduce the intensities of sun and storm. Because of the canopy's protection, few more serene habitats can be found on earth than the interior of a dense forest.

The understory

Smaller trees make up the understory. They may be young trees of the same kind as those that form the canopy, and some will eventually take their place among their elders; but many die without finding the opening they need. Or they may belong to entirely different species—low-growing trees such as dogwoods and hornbeams.

Sometimes the understory consists of young trees of a

species that is gradually replacing the predominant species. This happens in forests where the trees that make up the canopy cannot reproduce in the shade and in the leafy humus they have created.

In the forests of the North and mountainous West, for instance, young aspens do not grow well beneath their parent trees, but spruce and fir seedlings do. Consequently, in an aspen forest the understory is often composed of young coniferous trees. Usually the conifers grow so rapidly that in a few years they tower above the aspens, and the aspens die out. In this way the spruce–fir canopy replaces the aspen canopy.

Many birds and animals spend most of their lives in the understory. Creatures such as flying squirrels and many songbirds find the understory just the right height for nesting; the canopy protects their homes from hawks and owls and also from stormy weather, yet they are far enough above the forest floor to escape earthbound hunters.

The shrub layer

Shrubs are woody plants that have several stems. They form a distinct level in many forests; in others they are sparse or totally absent. A dense fir woods, for instance, where the ground is covered with dead needles, may have almost no undergrowth at all. Young broadleaf forests with more open canopies may be so thickly tangled with shrubs and vines that they are almost impassable.

Generally speaking, a particular kind of forest will have a particular kind of shrub growth—predominant trees are associated with predominant undergrowths. Thus in the beech–maple forests of the Northeast maple-leaved viburnum and witch-hobble are characteristic shrubs, and blueberry bushes grow frequently in many oak–pine forests. Mountain laurel and rhododendron are characteristic shrubs

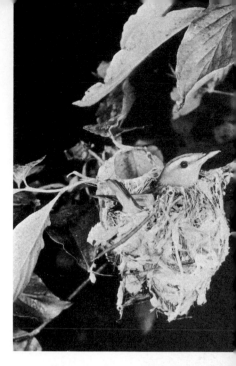

The red-eyed vireo constructs its nest five to twenty-five feet aboveground in the fork of an understory tree branch. Built of bark and grasses, vireo nests are often camouflaged with bits of spiders' cocoons, lichens, mosses, or flower petals.

Low-growing Pacific dogwoods make up the understory of this California incense-cedar grove in Yosemite National Park, California. These dogwoods flower abundantly in spring. In autumn, when the berries produced by the flowers are red, the leaves turn to brilliant shades of crimson and orange.

31

of southern mountain forests. In bottomland hardwood forests of the central eastern states, spicebush is a common shrub.

The shrub level rarely rises to a height of more than six or seven feet, and the thickest growth is often close to the ground. Shrubs provide protective cover for the activities of many small animals, such as shrews, deer mice, and chipmunks, which make their burrows near bushes and spend a good deal of time foraging in or under the overhanging branches. Some songbirds, including rose-breasted and black-headed grosbeaks, usually build their nests in thickets; certain ground-nesting species, such as grouse and ovenbirds, often locate their nests beneath the branches of shrubs.

The berries and seeds of shrubs are a good source of food, especially in autumn. Many kinds of insects feed in the shrub level, especially those associated with particular plants—the spicebush swallowtail, the witch-hazel cone-gall aphid, the laurel sphinx, and others.

The herb layer

According to technical usage, *herbs* are not merely the aromatic plants used in cooking but any green plants that have soft rather than woody stems. Most wild flowers and grasses are herbs. They make up the herb layer in the forest, along with other plants, such as ferns, mosses, and mushrooms, that grow close to the ground.

The exact nature of the herb layer is governed by various factors such as location, kind of soil, and amount of moisture. In moist swamp forests the ferns and mosses grow best, often luxuriantly. The drier forests of the western mountains and southern flatlands are more likely to support grasses.

The herb layer is usually most conspicuous in the spring. This is the time of beautiful woodland wild flowers. You can find anemones, spring beauties, trilliums, and Dutchman's breeches. Violets—purple, yellow, and white—abound and provide food for the caterpillars of the silver-spotted fritillary butterflies. In moist, lowland woods you can find heavy growths of rank-smelling skunk cabbage, which attract the carrion flies. Later in summer, when the canopy has closed overhead, most of the spring wild flowers die back to the soil. Only their unseen, underground parts live on.

Few forests have canopies so dense that no direct sun-

On the forest floor, the Kentucky warbler nests in a pile of leaves that blends so perfectly with the forest litter that predators prowling for eggs and nestlings often pass it by.

32

light comes through. Beams of sunlight fall through the openings in the canopy, forming sunflecks on the forest floor, and these sunflecks have great importance in the growth of the herb layer. Even though the sunflecks move across the forest floor quite rapidly as the sun moves across the sky, any plant in their path receives a "bath" of sunlight far more intense than light received during the rest of the day. In a few minutes such a plant may absorb more of the sun's energy than it otherwise gets in many hours.

Much of the characteristic appearance of the forest comes from the levels of the shrubs and herbs, even though these layers are dwarfed by the great trees growing above them. In a sense, the trees are too big: you don't see them. A woodland scene is normally composed of a few tree trunks, their tops cut off from view, and a variety of shrubs and herbs. We think of forests as open or dense according to the amount of low growth that blocks our view or impedes walking. And much of the most interesting life of the forest is found in the habitat of the ferns, grasses, and other herbs. Mice, insects, snakes, wood turtles, toads, such birds as veeries and hermit thrushes, and many other creatures live in the herb layer. Does hide their newborn fawns among the ferns; bobcats and foxes crouch behind logs and rocks to ambush their prey. In fact, not a single square inch of the herb layer is without some fascinating inhabitant.

Sword ferns (*above*) grow luxuriantly in the dank atmosphere of northwestern coastal rain forests. Reindeer moss (*below*), a lichen which is common in the tundra of the far North, may be found on the forest floor far to the south.

The forest floor

The forest floor is the wastebasket for all the layers above it.

Autumn is the time of the greatest accumulation, when the dead leaves of the hardwoods drift down abundantly, but all through the year there is a steady rain of other material—petals, fruits, seeds, bud scales, twigs, limbs, whole tree trunks, feathers, fur, animal carcasses, feces.

You can catch some of this material if you simply spread an old sheet on the ground and secure it with stones at the corners. Examine it each week—you will be surprised at what you find.

The amount you will collect depends a good deal on where you live. The heaviest annual leaf-fall, for instance, is in the low-lying equatorial jungles, where it averages some 10,000 pounds per acre. This amount gradually declines as one goes northward, until in subarctic regions the average may be only 800 pounds. Probably in most of the United States the average annual fall is 2000 to 3000 pounds.

All this material goes into the composition of the forest floor, which consists not only of this year's fall but of the fall from previous years that has not yet decomposed.

The total amount of plant and animal waste per acre is very large—as much as 194,000 pounds in fir forests where the earth is completely covered with slowly rotting needles. Even that is a dry weight, since scientists dry out the material before weighing it. In the forest this layer of decomposing matter absorbs a large quantity of water and retains much of it even during periods of dryness.

Scoop up a handful of this material. At first you will see little life in it, perhaps only an earthworm or a few black ants, but as you look more closely you will begin to see smaller creatures, mites and tiny spiders. A magnifying glass will open new worlds to you—pseudoscorpions, more mites,

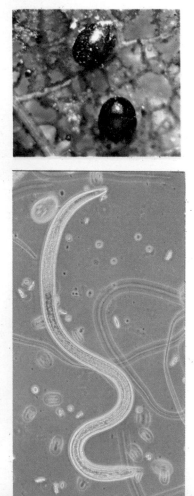

Soil organisms vary greatly in size. Some, such as earthworms and millipedes, are easy to see. Others, such as the mites *(top)*, here enlarged 22 times, can best be seen under a hand lens. But most soil animals, like the nematode *(bottom)*, here enlarged over 150 times, can be seen only through a microscope.

More than 2000 pounds of leaves, twigs, seeds, tree trunks, and branches may rain down each year upon an acre of forest floor. This debris, including the carcasses of dead animals, is ground up, chewed, dissolved, and eaten by millions of tiny animals and plants called decomposers. Thus basic substances such as carbon dioxide, water, and nitrogen are returned to the soil and air, and again become available to green plants for food-making.

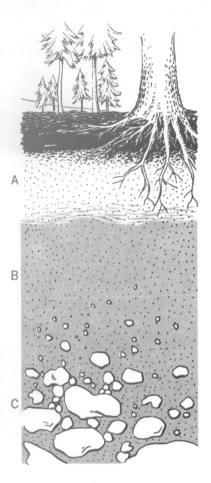

Soils have three major layers, or horizons. The A horizon, the uppermost layer, is commonly known as topsoil. Partly decomposed organic material is mixed with mineral particles in this horizon. Some of the fine clay and silt particles, as well as various organic and inorganic compounds, are washed downward into the B horizon, the subsoil below, by percolating water. The C horizon is the slightly altered bedrock that grades into unaltered material. In northern coniferous forests, partly decayed leaves accumulate on the surface, and little humus is mixed with the mineral soil. Iron compounds leach from the upper soil, leaving an ash-colored A horizon, and are carried downward by percolating acid waters. The B horizon is stained dark by iron deposits.

spiders, and tiny insects you didn't know existed. But the real surprise comes when you use a microscope. Now your view is crowded with strange little plant and animal forms. Indeed, most of the life of the forest floor is microscopic. The plants and animals on one acre may outnumber the entire human population of the earth by a million or more to one!

This mass of life attacks the forest's litter in many ways, releasing the wealth of energy and basic substances that is trapped in the decaying materials. Gradually these substances are broken down, recombined, moved about. Earthworms, burrowing animals, insect larvae, and other creatures tunnel continuously through the litter, gradually mixing it with the mineral matter of topsoil beneath.

Trees and soil

The soil of any forest is largely a product of its climate, its vegetation, and the underlying rock from which the soil has been formed.

In the deciduous forests of eastern North America, where the climate is moderate and where there is abundant moisture, the leaves of maples, beeches, oaks, hickories, and other broadleaf trees are quickly decomposed and mixed into the soil by teeming populations of small organisms. Here bacteria and fungi, springtails, millipedes, snails, and earthworms flourish. Usually leaves that fall in autumn are decomposed within three or four years, and some kinds may even disintegrate in a few weeks before the crop of a new autumn begins to drift down. Earthworms are particularly important in the mixing process because they literally eat the earth and everything in it. In their digestive tracts leaf fragments are crushed, partly dissolved, and mixed with rock particles. Soil known as *mull*, formed largely of earthworm casts and the excretions of other small animals, is a rich growth medium for many plants and is one of the reasons the floor of the eastern deciduous forest offers such an interesting variety of wild flowers.

Soils in the cold, moist forests of northern North America, where spruce and firs grow, are different. Decomposition of the acidic needles of these trees proceeds slowly and releases large amounts of acids. Water percolating through the thick accumulation of humus known as *mor* carries these acids down into the soil and leaches out many of the minerals. In fact, the soil just under the black humus layer may be light and sandy—as much as 80 per cent silica because the acids do not dissolve silica. The high acidity and low fertility of the soil, together with the thick humus mat on the surface, reduce the variety of soil organisms and smaller plants considerably. But in the coniferous forests of the Pacific Coast and on slopes of western mountains where the climate is milder, the needles of Douglas fir, ponderosa pine, and other trees decay more rapidly and produce less acid. The soils here are somewhat more like those of the eastern deciduous forests, and they may support a lush growth of herbs on the forest floor.

From year to year

Each year when you come back to the forest, it may seem unchanged. Seen from a distance, the tall trees of the canopy make a characteristic skyline. When you enter the woods, the lower trees of the understory and the shrubs beneath have the same appearance they had a year ago, or even five or ten years ago. You seldom find an unfamiliar plant or glimpse an animal you haven't seen before. This quiet, dimly lighted world gives an impression of calm stability and harmony. The forest seems eternal.

This is an illusion, of course. Changes are taking place in the forest all the time—some that you notice, many that you do not. From season to season you watch the hardwoods change; their airy spring foliage gives way to the cool green of summer, the fiery colors of autumn, and then to the stark bareness of winter. You are aware that many songbirds depart in autumn and return in spring, and that countless insects die each year and are replaced by countless others. There are daily changes too. The leaves of the forest trees manufacture food when sunlight strikes them, but in darkness photosynthesis stops. Many animals gather food all day, then retire at dusk; others sleep by day and hunt by night.

In spite of these periodic changes, at the end of a day, a night, a season, or a year, the forest is still there, looking much as it did before.

If you were to come back to the forest after an absence of a few years, however, the trees might be dead or gone. Where you had once stood in a great oak forest, enjoying the

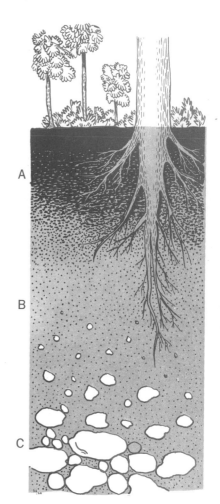

In eastern deciduous forests, leaves are decomposed rapidly and are mixed with the mineral soil by a myriad of small organisms. Percolating water carries some decomposed organic material deeper, but little of the iron compounds is washed downward. There is no sharp boundary between the A and B horizons.

coolness on a summer day, there might now be a ghost forest of dead trees and dry branches, or perhaps only an open field scorched by the sun. What happened to the forest that seemed so timeless and unconquerable?

The trees might have been killed by an invasion of insects; it has happened many, many times. In 1869, to take a recent example, a man living in Massachusetts decided he could produce a hardier silkworm than the Oriental species then used for making silk. He imported gypsy-moth eggs from Europe, thinking he would cross the gypsy moth with the silkworm moth. It was a hopeless idea from the start, as scientists know today, but the important point for us is that some of his gypsy moths escaped. Today, a century later, millions of their descendants have spread through our eastern forests, damaging and sometimes killing the trees over great areas by eating most or all of their leaves.

Perhaps the trees of your forest were killed by disease. The American chestnut tree, once a source of high-grade lumber and one of our most beautiful hardwoods, has nearly vanished. These trees were killed by the chestnut blight, a disease caused by a minute fungus. It probably came to the United States from Asia around 1900, carried on imported lumber. By 1904 the chestnut trees in New York state had begun to die, and by the late 1930s the blight had spread from Maine to Arkansas, killing almost every mature chestnut tree in North America.

Our native insects can devastate a forest too, even when they belong to the forest's natural community. Sometimes they become too numerous. Twenty years ago, for instance, the Engelmann spruce beetle, an insect that lives in the high spruce forests of Colorado, suddenly began to multiply. Before the beetle population fell back to normal levels, millions of spruce trees were dead, girdled by tunnels the beetles bored under the bark.

In addition to harmful insects and blights, nature brings violence to the forest—simple, crude violence. The changes may be dramatic; they almost always are. Hurricanes or

A collar of bristles encircles the grotesque head of the gypsy moth caterpillar. Brought from Europe by a naturalist in the mid-1800s, it accidentally escaped into the forest where it has become one of the most destructive of all insect pests.

Trees explode in flames as a roaring blaze sweeps through dense woodlands. Fires devastate about 3,000,000 acres of American forest each year, killing animals and plants and burning off the humus of the forest floor. Human carelessness causes 90 per cent of all forest fires.

38

tornados uproot huge trees and snap them like dry twigs. Forest fires, often started by lightning (or by careless visitors), rage through great tracts of forestland, reducing trees to charred skeletons, roasting wild animals alive, killing the living parts of the soil.

Sometimes volcanoes erupt, spilling hot lava on everything around them, forests and towns alike. In 1883 such an eruption tore apart an island called Krakatoa in Indonesia. Half the island vanished. The other half was covered with molten lava. It had been a rich tropical island, with thick forests of coconut palm, eucalyptus, and other trees, and with many flowering vines and colorful birds. A year later, when a scientist went to the island, he found nothing alive but one spider that had drifted in with the wind.

Will this be the fate of your forest—to perish in flames, to be ravished by insects or fungi, to be torn apart by earthquakes and volcanic eruptions? It could easily escape all these natural catastrophes only to be destroyed by human beings with bulldozers and chain saws.

The menace of man

Men, not insects or diseases or natural disasters, have been the chief agents of destruction in North American forests. This does not mean that men are necessarily to be blamed. The early settlers who needed food, timber, and living space had no choice but to cut the trees, burn the stumps, and plow the soil. But certainly men have often been unnecessarily wasteful of the wilderness, and many of their schemes could have been better planned and more intelligently carried out.

By the middle of the nineteenth century, thousands of acres of forest had been cleared in New England to make way for farms—yet in the period since the Civil War much of this acreage has been abandoned. There were many reasons, but a major one was the failure of crops. The soil had

Lashing logs together with chains, a lumberman prepares newly cut timber for the trip to a sawmill. Since 1900, when forest conservationists and lumbermen began to develop sound programs for reforestation, an increasing portion of the 460 million acres of commercial forestland has been maintained by replanting and controlled cutting.

The end and the beginning of a forest are dramatically symbolized by this charred tree trunk and the clump of grass growing at its base. Grasses are often among the first plants to appear after a fire. Their deep-growing roots knit the soil together and help to prevent erosion.

Once heavily forested, Copper Basin in Tennessee is today a vast wasteland. Many trees were cut down to supply fuel for copper smelters, such as the one at Ducktown *(upper right)*. Eventually acid fumes from the smelters killed the remaining trees and all the shrubs and herbs as well. Attempts to reforest this area have failed because rain has washed away the topsoil.

been ruined by too much cultivation without proper fertilizer or crop rotation; men had taken whatever they could get from the earth without giving anything back. The settlers sought more fertile land in the West, moving ever onward, destroying the forests wherever they went. Many settled in Ohio, and the forests there dwindled. In 1788 Ohio was 95 per cent forest. In 1853 it was still 54 per cent forest. But by 1940 the forestland had shrunk to 14 per cent.

What happens to an area after it has been cleared of forest, cultivated, and then abandoned? Once destroyed, are the forests gone forever?

Suppose someone has decided to clear a large area of oak forest for cultivation. Men with chain saws rip through the huge trunks and topple them. Next they haul the logs away on trailer trucks. Then they dynamite the large stumps and scrape away the smaller stumps, understory trees, and shrubs with bulldozers. Finally, with great tractors and many-bladed plows, the men churn up the precious humus of the forest floor.

The years of cultivation

Now the farmer plants his crop in the field where the oaks once stood. By August long rows of corn line the field, close-set and unyielding like an army in mass formation. Nothing else appears to be alive.

But look again. There are weeds—the "wild" plants—growing between the stalks of corn. Where have they come from? Some have sprung from seeds that were left in the ground, others from seeds blown in by the wind. And where there are plants, there are animals. Insects are feeding, birds are swooping down to catch the insects. (The birds are chiefly mealtime visitors who nest in nearby wooded or bushy areas.) In the evening rabbits, raccoons, and deer come to nibble the corn. Most of these animals, too, live in the shady woods and come to the field only to eat. But by autumn the meadow mice have begun to make their homes in the field, and a few woodchucks (groundhogs) have dug burrows around the edge.

For several years the farmer cultivates his land; then, perhaps discouraged by poor yields, he abandons it, and it becomes an old field. What will this old field look like in twenty, thirty, fifty years? Will it remain an open meadow, ragged and weedy, with neither the fullness and dignity of a forest nor the neatness and regularity of a farm?

In the slanting sunrays of an autumn afternoon, a raccoon feeds on corn. Its cornfield raids are more frequent in the summer, when the kernels are juicy and tender. But if nuts and berries or crayfishes, frogs, and other small creatures are scarce in autumn, it readily eats mature corn.

The first year after cultivation

Suppose you could watch a motion picture of this field filmed at intervals over a period of many decades. You would see how the old field changes.

The first sequence would show the field in the spring of the first year after it had been abandoned. The soil is covered only by corn stubble from last year's crop and by brilliant, yellow-flowered wild mustards. The hot sun falls directly on the earth between the plants. By midsummer, however, the field is covered with a tangle of stems, leaves, and small yellow and white flowers. These are annuals—horseweed, chickweed, pigweed, ragweed—plants that live only one season but drop seeds that sprout the following year. These dense weeds shade the soil a little, reduce the force of the heavy rains, deflect strong winds. The *microclimate* (little climate) has changed at soil level. It is cooler and more stable; the erosion of the earth has been slowed by thousands of miles of roots spreading through every acre.

With a milder microclimate and a greater variety of plants, many new animals come to feed in the field. Grasshoppers leap among the weeds, leaf hoppers suck the juices from plants, aphids multiply. And now come their predators, the perfectly camouflaged crab spiders, praying mantises, spotted red ladybugs, wasps, hornets, robber flies, and ground beetles.

Next come more birds. Some perch on dead cornstalks, then flit out to snatch flying insects from the air; others hop on the ground, finding insects in the litter of dead plants and on low-growing stems. In late summer, sparrows, doves, goldfinches, quail, and pheasants fly in to feed on ripening weed seeds and fruits. Many of the birds are visitors from the woods or scrublands. Brown-and-yellow meadowlarks, however, nest on the old field; so do killdeer and vesper sparrows with tails striped like rocket fins. Both species build their nests on the ground, in the weeds and grass.

Other animals, too, are now moving into the old field in

increasing numbers. Meadow mice forage voraciously on grasses and wild flowers; rabbits come out at dusk to nibble the clover; shrews are multiplying rapidly; moles are tunneling among the roots. With these animals come the larger predators. By day the hawks soar and swoop, clutching mice in their powerful talons; many of the hawks are thick-set and stocky—red-tailed hawks, red-shouldered hawks, broad-winged hawks, and rough-legged hawks—but there may also be smaller ones—sharp-shinned hawks and sparrow hawks. At night the owls take over, dropping on silent wings to snatch mice and rabbits from the grass. Chiefly they are barn owls, great horned owls, and barred owls, but nearly all species of owls prey on meadow mice.

Now a new predator appears in the old field, a satiny black creature slithering among the weeds: a six-foot pilot snake. (As a matter of fact, the old field is the habitat of many of our common snakes—garter snakes, ribbon snakes, milk snakes, and others.)

The seeds of autumn

When autumn and winter approach, the dying and dead old-field weeds are broken by wind and bent by rain. The ground is littered with stalks and leaves, and with the carcasses of countless insects. Here is rich food indeed for fungi and scavengers, a feast for springtails, slugs, and beetles.

Most important, millions of seeds are falling to the soil. Many are eaten by birds and mice; others fall into cracks and are washed down too deep to sprout; still others are rotted by fungi, chewed by insects, or swallowed by earthworms.

But this great loss is overcome by an even greater productivity. Counts have shown as many as 160 million seeds in the top four inches of an acre of farmland. A dandelion plant can produce 12,000 seeds in one summer; a pigweed six million! No wonder there are always enough seeds left

SEEDS ON THE MOVE

Seeds are transported in various ways from their parent trees, shrubs, and smaller plants to new growing sites. Many kinds of seeds, especially those of grasses and small herbs, simply fall to the ground. Some are carried through the air by wind. Some cling to the fur of animals. Others shoot from their pods like tiny artillery barrages.

Only a small percentage of the billions of seeds produced annually grow into new plants. Many fall upon bare rock or other spots unfavorable for germination. Many are eaten by birds, rodents, and insects. Indeed, it is estimated that 200 to 400 tree seeds must reach the ground for each seedling that becomes successfully established.

Seeds as hitchhikers

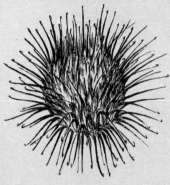

BURDOCK

Animals, including man, are effective agents of seed dispersal. Birds, for example, eat fleshy fruits and pass seeds through their bodies unharmed. Squirrels collect acorns and bury them; although they dig up most of the nuts later, they often overlook some. Many mammals carry seeds on their furry coats, especially sticky seeds such as those of mistletoe or seeds with hooks or barbs such as those of burdock, beggar's tick, sticktight, and cocklebur.

Seeds as bullets

The seedpods of some plants contract as they dry, ejecting their contents as "bullets" over surprising distances. Among such plants are jewelweed, violet, witch hazel, and touch-me-not.

WITCH HAZEL

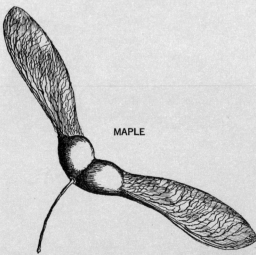

MAPLE

Seeds as gliders

Many kinds of seeds are transported by the wind. When tumbleweed seeds mature, the entire plant dries up and rolls away in the wind, scattering seeds as it goes. The seeds of poppy and iris are shaken from dry pods when the wind blows. Ash, pine, and maple seeds travel great distances on "wings." Still other seeds glide through the air on feathery plumes, fuzzy coatings, and even parachutes like the goatsbeard's (*opposite*).

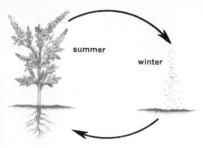

Annual plants live for one year, produce seeds, and then die. New plants develop from seeds the following year.

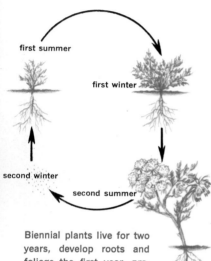

Biennial plants live for two years, develop roots and foliage the first year, produce seeds the second year, and then die.

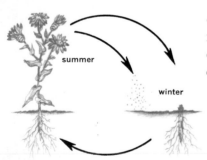

Perennial herbaceous plants often live several years. Many die back to the roots during the winter. Most produce seeds every year.

over to produce next year's weed crop. Yet even while millions of annual seeds are raining down on the old field, a host of longer-lived plants—the biennials and perennials—are beginning to grow in the field.

Goldenrods and asters move in

Now our film of the old field moves ahead to the next spring and summer. New plants appear. Some are coarse grasses. Some are leafy plants with tall stems and clusters of flowers, some yellow, others pink or rose. These are the biennials and perennials, plants that were seedlings the year before. Biennials, such as Queen Anne's lace, common mullein, and moth mullein, take two years to complete their life cycle from seed to seed. Perennials live several years and produce new seeds each year.

Often the perennial herbs die down to their roots each winter, as if they were annuals, but next spring new stems and leaves spring up from the same roots. The basal leaves of most perennials, especially in the South, stay green all winter. Perennials compete with the annual weeds that flourished during the first year of the old field, and by the fourth or fifth year they take over almost entirely. Then, with more plants, there are more animals. Several kinds of meadow-nesting birds now appear—savanna sparrows, bobolinks, red-winged blackbirds. In western areas there may be sharp-tailed grouse and greater prairie chickens. More hawks appear overhead. Foxes and weasels prowl through the dense growth in search of prey.

A great variety of green plants now seem to occupy every inch of the old field, though in most areas the goldenrods and asters predominate. Milkweed, black-eyed Susan, oxeye daisy, Queen Anne's lace, broom sedge, and many others can be found. Unlike the wild flowers of the woods,

Goldenrods are a source of food for a host of insects, including many kinds of bees, flies, aphids, wasps, and beetles. Here a black blister beetle (*left foreground*) feeds upon pollen-laden flowers, while a well-camouflaged orange ambush bug (*right foreground*) waits for an unwary insect to stray near its powerful forelegs. Yellow-striped beetles swarm over the flowers. Like several other species of insects, they eat only goldenrods.

48

most of these are summer-blooming species, and some, such as the asters, keep their petals almost until the first snowfall.

Many insects now inhabit the old field, fluttering around the bright flowers—bees, wasps, yellow jackets, flies, butterflies, moths. And there are strictly predatory insects, too, including the ambush bug. This rapacious insect waits for other insects to visit blossoms for nectar and pollen, fastens its clawlike legs around its victim, and then sucks out the body fluids. A similar insect is the assassin bug, which can inflict a painful bite with its curved, jointed beak. Spiders, however, from the small crab spider to the big orb-web weavers, are the most numerous and the most important predators of the old field.

Caterpillars, which are the larvae of butterflies and moths, are also commonly found on asters and on goldenrods—or, for that matter, on almost any green plant. Some are furry and beautifully marked, others are naked and, from the human point of view, ugly. Most caterpillars are vegetarians and all are ravenous feeders. The large ones eat mostly leaves. Smaller ones often bore into stems, and some spend part of their larval stage inside galls on stems, leaves, or roots.

A gall forms when the females of certain insects lay eggs in living plant tissue. When the eggs hatch, the larvae live for a time inside the gall, eating the plant tissue as they grow; then they bore holes to the outside and emerge. The gall itself is an abnormal growth of the plant, the plant's reaction to irritating fluids injected by the female insect when she lays her eggs and released by the developing larvae. Each of the hundreds of kinds of gall wasps, gall flies, gall aphids, and gall mites causes a gall of distinctive shape, and with a few exceptions each lives always in the same species of plant. Thus the plant and the insect live in a close relationship, beneficial to the insect and in most cases not harmful to the plant. Goldenrods often support several kinds of galls. Oaks, maples, aspens, and willows are trees commonly afflicted with galls.

The trees take over

Our old field film is moving on now to the fifth or sixth year, and you can catch a hint of what is to come. Here and there small trees and shrubs are beginning to become conspicuous among the perennial herbs. By the time our film passes the tenth year, they have grown taller and broader, and there are more of them. Goldenrods and asters are still kings of the old field, but another "battle of plants" is beginning.

It may seem that the weeds, with their millions of seeds, would be unbeatable. But the woody trees and shrubs have the advantage. When the goldenrods and asters die down to their roots each autumn, the trees remain standing and thus they have a head start in the spring. As their hard trunks and branches fill out, they become more and more predominant in the landscape.

Gradually great changes come to the old field, as the years go by and our film speeds on. The level of vegetation rises; distinct layers begin to form. Some of the goldenrods and asters that once towered over the old field now are in the shade of trees. Cut off from sunlight, deprived of the water and minerals that the tree roots absorb, and buried by dead leaves, the weeds are beginning to disappear.

Soon the young trees are thick. Their polelike trunks shoot higher and higher—twenty feet, thirty feet—suggesting the forest that is to come. But they are a long way from a forest yet, especially the forest that once occupied the old field, for most of these new trees are conifers—pines or cedars—not the oaks and hickories whose broad leaves formed a canopy over this land in years gone by.

How long will it be before a canopy forms? On a Long

Island old field the change requires about half a century. In southern New Jersey, where scrub pines are the predominant old-field trees, it may take only twenty years. But in Missouri, fields forty years old still look like brushland, and one field in California after the same length of time had only one or two tree seedlings per acre.

Why do trees grow quickly in one region, slowly in another, hardly at all in a third? The answer depends on many factors: temperature, rainfall, wind, soil, elevation above sea level, the kinds of plants and animals present, to name only a few. Indeed, there are so many factors that scientists are only now beginning to understand the principles governing the growth of natural plant communities. Compared to calculating all the factors that influence the growth of just one plant in our old field, plotting the flight of a moon rocket is simple arithmetic.

The pine woods

The old field is now almost entirely covered with pines. Though the trees are still young, the area is beginning to take on the appearance of a forest; and thus the weed field is disappearing. As it vanishes, the varied animal life of the field goes with it and is replaced by the animal life of the woods.

Meadow mice, which thrive on weeds and grasses and make their nests of grass, are replaced by white-footed mice, which eat the seeds of trees and shrubs and make their nests in hollow logs, stumps, and tree holes, in burrows, or among the branches of the understory. They are good climbers and can seek their food high in the trees as well as on the ground.

As the pine wood thickens, deer roam through, hidden by the dense growths of needled branches. The field birds disappear and are replaced by birds that customarily live on the edges of forests: rufous-sided towhees, cardinals, various flycatchers, yellow-breasted chats, and many other species of warblers. As the pines grow taller and taller, some of the deep-woods species appear: woodpeckers, blue jays, nuthatches.

Dead needles cascade to the ground and form circles beneath each tree. As the trees grow taller these circles become wider until eventually they merge to create a continuous forest floor. Finally the crowns of the trees close over the ground and form a dense canopy. Nearly all the old-field herbs have disappeared, and the floor is dimly lighted. Here and there, however, a few plants that grew here long ago, when the oak forest covered the land, begin to reappear—spotted wintergreen with its waxy white flowers and mottled leaves; velvety mosses; scaly lichens.

Pine trees now cover the entire field and grow so close together that the needles on their lower branches turn brown from lack of sunlight. Soon the smaller pines begin to die.

The oaks and hickories return

As the camera scans the forest floor, you see very few pine seedlings. Apparently pines do not grow well beneath their parent trees. Other seedlings do better—young oaks, black cherries, hickories. Many die before they have grown more than a few inches, but others survive and grow taller.

Acorns, cherrystones, and hickory nuts are much heavier than pine seeds; usually they fall close to their parent trees. These new seedlings, then, must have been "planted" by animals. Squirrels, chipmunks, and blue jays probably carried the acorns and other seeds from a nearby forest. The cherrystones may have been dropped by starlings that came to roost among the pines, or they may have been dropped by a fox that strayed through the woods. As the animals digested their meal of fruit, the stones passed unharmed through their digestive systems and fell with the droppings to the ground.

The film speeds on. The oaks and hickories gradually rise to form a dense understory. As the pines mature, they also weaken, and in each windstorm a few topple against their neighbors or fall directly to the ground. Here and there young oaks and hickories begin to form patches of broad-leaf canopy.

As more pines die, oaks and hickories take their places. A storm—a hurricane—helps, snapping off hundreds of mature pines. In this event, some pine seedlings may spring up and flourish for a few years in the open spaces, but in a few decades they reach maturity, die, and are replaced by broadleaf trees. Finally only a few scattered pines remain.

On the forest floor humus is accumulating again. More and more dead oak and hickory leaves are mixed with the dead needles and with the carcasses and excretions of mam-

The aspen trees on this Arizona hillside are growing where a spruce–fir forest once stood. A fire killed the coniferous trees many years ago, and now only a few of their fallen, decaying trunks remain. Not long after the fire, aspens sprouted from the roots of stunted parent trees that had been unable to attain tree size in the original forest. Without competition from the formerly predominant firs and spruces, these young broadleaf trees grew very quickly. More and more root sprouts shot up, until the hillside was covered with aspen trees. Now, years after the fire, spruce seedlings are coming up beneath the aspens. In time, as these evergreens grow larger, the aspens will begin to die. They will not all die, however, and their roots will be able to send up thousands of sprouts should fire destroy the new coniferous forest.

mals and birds and insects. A rich deposit is building up for the bacteria, fungi, and earthworms to work on.

More variety can be noted in the plants of the forest floor. You see hair-cap moss, ground pine, liverworts, and here and there the bright-red British soldier lichens. At the same time a shrub layer of huckleberry, wild raspberry, and mountain laurel begins to form.

Many generations must pass, even hundreds of years, before the young broadleaf trees will equal the trees of the great forest that once covered this land. But clearly the oak–hickory forest is returning.

Succession

This stone wall, which once separated plowed fields, is striking evidence of the forest's ability to regenerate itself. Such walls are common throughout the northeastern states where thousands of acres of forested land were cleared for cultivation in the nineteenth century and then abandoned.

Our film is ending; the final credits unroll. They show that the motion picture was produced in one of the mid-Atlantic states, perhaps New Jersey or Maryland or Virginia. Had it been filmed elsewhere, the cast would have been different but the plot would have been the same.

It is an ancient story, the oldest in nature. One group grows successfully for a time, but eventually gives way to a new group that is better suited to changed conditions. Eventually a relatively stable balance is reached; but only

Undaunted by gasoline fumes, incinerator smoke, and poor soil, weeds grow luxuriantly in a vacant city lot. Left undisturbed, the lot may eventually be overgrown by such city-tolerant trees as ailanthus, sycamore, American ash, or horse chestnut.

until disaster strikes—a storm, a fire, the chain saw and bulldozer—and then the process begins again.

Scientists would entitle our film *Succession: The Gradual Replacement of One Community by Another.*

Succession means a shift in plant and animal populations. Shrubs and trees invade a weed field. One kind of forest replaces another. In each case the landscape changes strikingly.

A succession of plant communities always is accompanied by a succession of animal associations. Plant succession leads the way because plants are the foundation of every food chain. The species of plants in an area always determine the species of plant-eaters, and these in turn largely govern the species of predators in the same area.

This fact may account in part for gradual extensions of the ranges of some animal species. The cardinal used to be considered a southern bird; now it lives as far north as Maine and southern Ontario. There are a number of reasons for this movement, including a recent warming trend in the northern parts of the continent. But certainly the increase of areas covered by young forests—as northern farms have been abandoned—has greatly expanded the kind of habitat utilized by cardinals.

Succession is universal, but it may proceed very rapidly or very slowly, depending upon the circumstances. Generally, changes take place quickly in the tropics and in moist, temperate areas but only very slowly in cold or dry areas, as in tundra or desert. And the changes within a

59

given succession do not occur with equal rapidity. In the development of our oak forest, for example, annual weeds were replaced by perennials in only four or five years, and perennials gave way before pines in another decade. The final change from pines to oaks and hickories, however, required a century or more.

How can we study succession?

Our imaginary film of forest succession on an old field covers a period of time that could easily exceed your lifetime. It might cover several centuries, or as much as a thousand years. Then how can we study such gradual changes?

We can study prolonged successions by piecing together the evidence from different areas. Different oak forests show different stages of development; by putting them together a trained observer can draw various conclusions about the development of an oak forest from old field to mature forest.

In certain circumstances you can find an entire succession laid out in a limited area. This happens when a pond or lake gradually dries up. As the water level falls and the pond shrinks, vegetation moves inward toward the receding shoreline. As pickerelweed and spatterdock at the water's edge close in around the shrinking pond, perhaps sedges occupy the area these plants formerly covered. Where sedges grew only a few years ago, now there is a shrubby leatherleaf thicket. Farther back from the pond margin are successive bands of highbush blueberries, red maples, pines, and oaks.

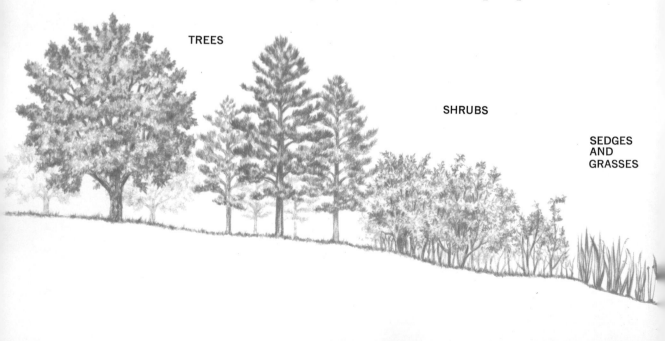

TREES

SHRUBS

SEDGES
AND
GRASSES

Today, the place where the shoreline stood five hundred years ago is pine forest; and where pines grew then, now there are oaks. Between the oaks and the present shoreline you can retrace the steps in the succession that probably led to the present broadleaf forest, and you can picture how the forest will eventually cover the pond.

By such observations scientists can puzzle out what has happened in the past on a given piece of land, and what is likely to happen in the future. And you yourself can do it too, at least up to a point. When you see an abandoned field, you can identify the perennial weeds that have become predominant in it, and perhaps you can find scattered shrubs or tree seedlings, already foreshadowing the change that will come as the open meadow changes to thicket, to young woodland, and ultimately to mature forest.

Impediments to succession

To the working ecologist the entire order of the communities that follow one another in a natural succession is known as a *sere*; the final community in the sere is known as the *climax.*

For example, the entire progress from newly abandoned field to oak–hickory forest is one sere, and the oak–hickory forest itself is the climax. Each phase of the sere is called a *seral stage.*

But in some respects the term *climax* is ill-chosen. It implies permanence and stability, although ecologists can never be sure whether or not a particular forest has reached

Ponds begin to disappear as soon as they are formed. Wind and rain carry in soil, dead plants, and animal carcasses. Slowly the bottom of the pond is filled with debris, and the water becomes more shallow. Because pond plants are adapted for growing in certain depths of water, distinct circles of vegetation surround the open water in the center of the pond. Some plants grow only underwater; some have floating leaves; others grow up out of the water. Still others grow only at the swampy edges of the pond. As the water depth continues to decrease, underwater plants advance across the bottom and in time completely cover it. Then floating plants grow inward, followed by plants that extend above water, such as grasses and cattails. Just behind these plants are shrubs and trees which eventually overgrow the old pond site and form a new forest.

SWAMP
PLANTS

FLOATING
ROOTED
PLANTS

UNDER-
WATER
ROOTED
PLANTS

FLOATING
PLANTS

OPEN
WATER

a state of balance. Even forests that appear to have changed little for a thousand years or more may be undergoing new phases of development. Leaving aside the disasters that can destroy them, there may still be subtle changes in the forest community—changes in rainfall, temperature, soil composition, animal population. After a long enough time, any of these things could cause a change in the most advanced forest.

We therefore use the term *climax* with the understanding that it is a handy way of referring to the most stable and highly organized stage toward which seres in a particular region appear to be moving. It is at best a relative term. No natural community ever achieves a state of perfect equilibrium.

A very great number of ecological problems, large and small, remain unsolved. Occasionally it seems as if each time scientists learn something new they merely reveal a larger area of our ignorance. Many more trained ecologists and field workers are needed, of course, but professional scientists are not the only ones who contribute to the progress of ecology. Amateurs can help, once they have learned to observe nature intelligently.

Many factors influence succession, and many things can interfere with it. We know that animals play an important role which can be either a help or a hindrance. Scientists once studied the acorn crop of a mature oak tree in North Carolina. They found that the tree produced 15,000 acorns. Of these, 83 per cent were eaten by deer and other mammals, 6 per cent were infested by weevils and moth larvae, and nearly all the rest were naturally sterile or deformed. Only eight tenths of 1 per cent of the acorns sprouted; half of these died as seedlings.

But acorns are by no means the only seeds affected. Deer, bear, foxes, mice, grouse, and many other birds and mammals all feed heavily on plant seeds at one season or other, thus altering or retarding the progress of a sere. For instance, some seeds are more appealing than others. Black and red oaks have bitter acorns, white oaks have sweeter ones, with the result that black- and red-oak acorns are not eaten by larger animals so readily as white-oak acorns and have a better chance of germinating successfully.

Deer can change the makeup of a forest if they are numerous enough, for they browse more heavily on some trees than on others. Red maple, tulip tree, crab apple, horn-

Like many other forest animals, the white-footed mouse often eats acorns. In years of bumper acorn crops, one mouse may store away hundreds and eat them during the winter.

beam, sassafras, and flowering dogwood are among their favorites. As a result, other species, which the deer do not eat so readily, have an advantage.

In some regions porcupines can exert a considerable influence. One mature porcupine, by eating a circle of bark around the trunks of trees, can kill a large number in his lifetime. In New England porcupines have destroyed whole groves of sugar maple, and they attack beech and hemlock, too. In the same area other trees may be left untouched and thus may become the predominant trees, at least for a time.

But man himself has done the most to interfere with succession. By clearing the forests he destroys climax growths. By plowing the soil he mixes the layers of humus and destroys the horizontal banding of soil which has taken centuries to form. With pesticides he eliminates useful insects along with harmful ones. The poisons may even be passed along forest food chains and affect birds, mammals, and fishes. Sometimes he purposely stops succession at a particular seral stage. In the South, for instance, yellow pines form an early seral stage; but because they are commercially valuable trees many forests are burned over with controlled fire to prevent the growth of less valuable oaks and hickories, which otherwise would replace the pines.

Deliberate, or prescribed, burning may save one of our rarest songbirds, the Kirtland's warbler. The warbler's peculiar nesting habits led very nearly to its extinction not long ago. It nests only in a limited area of Michigan, on the ground near or under the shelter of young jack pines that are no shorter than five feet and no taller than eighteen. When the pines have grown to a height of much more than eighteen feet, these warblers will no longer nest under them.

In former years, when forest fires were common in Michigan, jack-pine seedlings often sprang up in the burned-over areas, furnishing many nesting sites for the warblers. But with the practice of modern fire prevention and control, the number of burned-over areas has greatly declined, thus reducing the thickets of young jack pine and suitable nesting sites for the warblers. The population of warblers has fallen sharply. Today, by practicing prescribed burning and selective planting, conservation authorities in Michigan are attempting to hold back the succession and encourage the warblers to nest, with the hope that the population of Kirtland's warblers will increase from its present number of about a thousand.

Kirtland's warblers, such as this handsome male *(top)*, depend upon forest fires for their very existence. These rare songbirds nest only in young jack-pine forests, which spring up on burned-over land *(bottom)*. Once natural forest fires continually created such nesting sites, but now men control fires so effectively that there are fewer areas covered by young jack pines, and as a result, few Kirtland's warblers. Conservationists in Michigan are intentionally burning selected areas to provide nesting sites for the estimated 1000 remaining birds.

A lone mule deer bounds through a forest clearing on Arizona's Kaibab Plateau. In 1907, the deer on the Plateau numbered about 4000. To protect the herds, government authorities prohibited deer hunting and employed professional hunters to kill off the deer's natural enemies. The results of this policy were disastrous. By 1924 the herd had increased to an estimated 100,000 animals, many thousands more than the area could support. During the next two winters 60 per cent of the deer starved to death, but not before nearly all the herbs, shrubs, tree seedlings, and lower branches of mature trees had been eaten or badly damaged. Today, there are only about 10,000 deer on the Plateau, and hunters can shoot them in season. Thus the deer population is kept within the limits of its food supply, and the 700,000-acre Kaibab forest, which eventually might have been turned into a tree-dotted grassland by overbrowsing, is regaining its former beauty.

The menace of erosion

Scores, even hundreds, of years may be required to repair the damage done in a single locality by insect plagues, by foraging mammals, by natural disasters such as forest fires, by men who cut down the trees and cultivate the land. But we know that so long as soil suitable for plant growth remains, the damage will be repaired.

The damage done in many areas by soil erosion is, however, another case; it may require millenniums to erase. When wild plants are destroyed, the dense tangle of roots that has held the soil intact for thousands of years is suddenly removed, and the soil may easily be carried away by wind and rain.

The depth of fertile soil on most of the earth's surface is no more than a few inches—a few feet at most. Below that are sterile sands, gravels, clays, bedrock—far less suitable for plant growth. In many places where the wilderness has been cleared and the land plowed, fertile soil can be washed away by rainstorms in a few years or decades. This is precisely what has happened in far too many areas; only barren clay, sand, or rock remains.

During the flood seasons in many parts of the world, great rivers are brown with mud—fertile soil being carried away to the sea. Where huge dams are built to control the floods, the lakes behind them may fill so rapidly with silt and mud that the dams will be useless before this century has ended. Too little attention is paid to the condition of land above the dams, and the reckless practice of farming steep slopes and riverbank areas is often allowed to continue. How much wiser it would be to encourage forests to grow on these lands, so they can serve as natural filters and as erosion controls!

Held together by the roots of a blackjack oak, a ten-foot pillar of earth remains to show the former surface level of a badly eroded area in Tennessee. Stranded trees usually die from damage to their exposed roots, lack of water, or windfall.

A forest fire devastated this portion of Coeur d'Alene National Forest in Idaho years ago. All the forest trees, shrubs, and herbs were killed, and the protective mat of decaying plant materials was burned off the soil. Without living roots to hold it and without cover to deflect the full force of wind and rain, the topsoil began to erode. If foresters had not quickly replanted the area, it might have eroded seriously enough to have remained barren for hundreds, or even thousands, of years.

Succession within succession

The living community of plants and animals in even the smallest microhabitat is extremely complex. To study thoroughly the succession of different plant and animal species in a single tree hole, for instance, a scientist may need to make continuous observations over a period of several years or more. Every important event must be noted. Every new relationship must be measured. All major environmental changes—changes in temperature, moisture, and the like—must be detected.

Thus thorough documentation of the changes in even one microhabitat may amount to an enormous mass of data. All of this must be tabulated and correlated before any conclusions about the causes of the succession in the tree hole can be reached. Just to summarize the data, a scientist might have to write a report much longer than this book.

If this is the case with a tree hole, think how much more complicated the succession in an entire forest is. We have spoken of the simple progress from old field to mature forest, but within this succession hundreds of others are occurring too—successions in tree holes, in fallen logs, in ponds and brooks, even in abandoned birds' nests and in acorns. In many cases the effect of very small successions seems of little importance; yet an effect is there. The relationships in a particular forest between all the phases of all successions, major and minor, are so complex that it would be scarcely possible to discover them in one lifetime.

Of course, the principles are important too, and fortunately they can be discussed by using simplified examples, such as our succession from an old field to an oak–hickory forest. But you must bear in mind that a forest is more than a collection of abstract principles. It is, for that matter, more than a collection of trees. It is an incredibly intricate, constantly changing community of plants and animals. If it were not, it would be neither beautiful nor interesting—nor alive.

Slowly crumbling in decay, the stump of a forest tree provides food for plants and animals in the eternal process by which dead things are returned to the soil that nurtured them.

Seasons in the Forest

SOME changes that occur in the forest are violent and unpredictable. Others are calm and regular: these are phases of the great natural cycles that take place daily, seasonally, yearly, or at even longer intervals.

Simple explanations used to be offered for these changes. People believed, for instance, that plants flower in spring so that their seeds will be ready for sowing in autumn; that squirrels collect nuts in autumn to provide a store of food during the snowbound months of winter. But now scientists realize that although these may be the end results, neither plants nor animals behave as they do for these reasons—or indeed for any "reasons" at all.

Nature does not think or plan ahead. Instead, all living things are regulated by automatic inner processes phased so closely to outer events that they are called "biological clocks." Some plants and animals respond to light, particularly to changes in the number of hours of daylight from season to season. Some respond to changes of temperature or moisture. But many of the factors that determine the rhythmic activities of plants and animals are still unknown.

Of all the rhythmic changes that take place in temperate

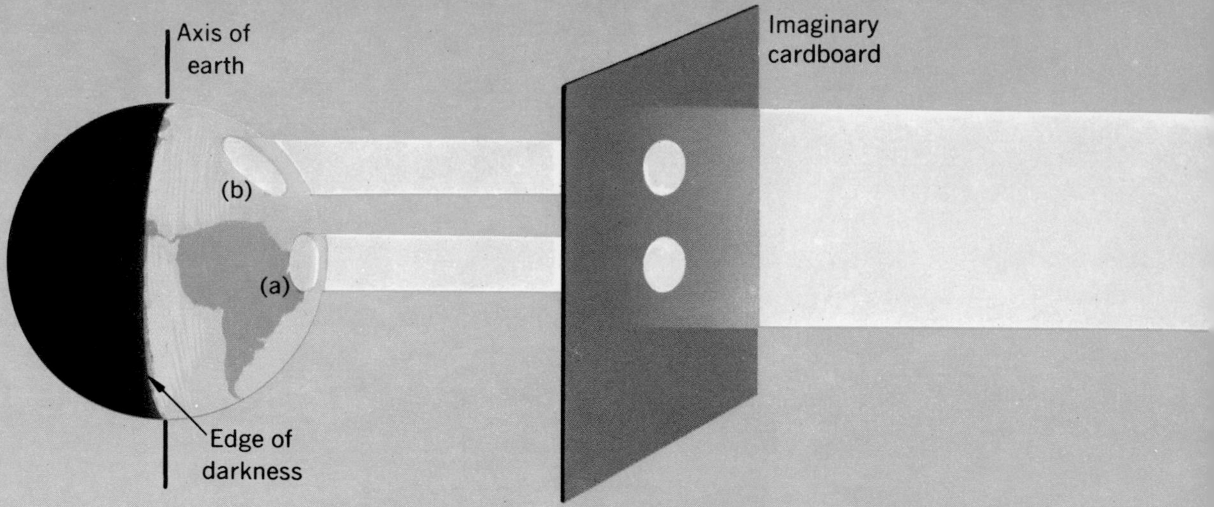

Axis of earth

Imaginary cardboard

(b)

(a)

Edge of darkness

LONG DAYS AND SHORT DAYS

In the summer when the North Pole is tilted toward the sun *(opposite)* it is in sunlight 24 hours a day. In winter there are no hours of daylight. Because the Northern Hemisphere is covered with more sunlight during the summer, the days are longer. In winter when the Northern Hemisphere is covered with more darkness the days are shorter.

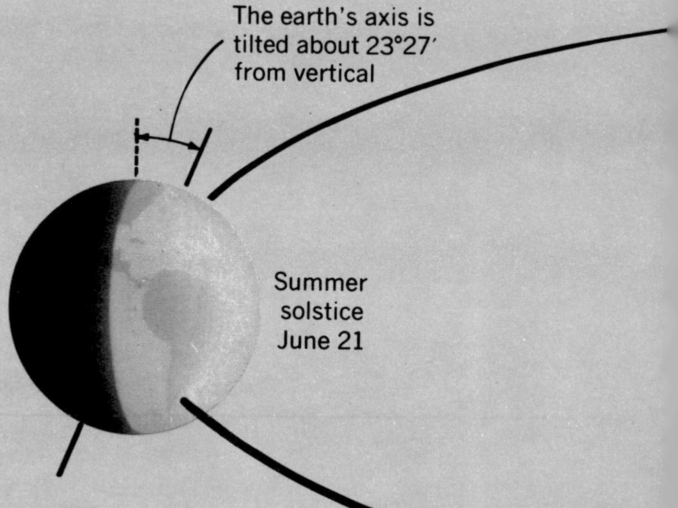

The earth's axis is tilted about 23°27′ from vertical

Summer solstice June 21

THE CHANGING OF THE SEASONS

The changing seasons are the result of the permanent 23½ degree tilt of the earth's axis. Imagine a huge sheet of cardboard with two holes of the same size cut through it. Sunlight passing through the holes falls on the earth *(top diagram, opposite)* at the equator (a) and in the Northern Hemisphere (b). Since at point (a) the sunlight strikes the earth vertically, the sun's energy is concentrated, and temperatures are higher. At point (b) the sunlight strikes the earth at a slant, and, therefore, the sun's energy is spread out and temperatures are lower. In the bottom diagram *(opposite)* the North Pole is pointed toward the sun. When this happens the concentrated sunlight at the equator shifts northward and the temperature rises bringing summer to the Northern Hemisphere.

In the illustration below the four seasons are shown as the earth makes its yearly orbit around the sun. Notice how the North Pole tilts away from the sun in winter causing the temperature to drop. Just the opposite happens in the Southern Hemisphere where summer temperatures occur during the northern winter. In the spring and again in autumn the poles point neither toward nor away from the sun and the sunlight is evenly distributed both north and south of the equator. Thus, the temperatures of these seasons are moderate.

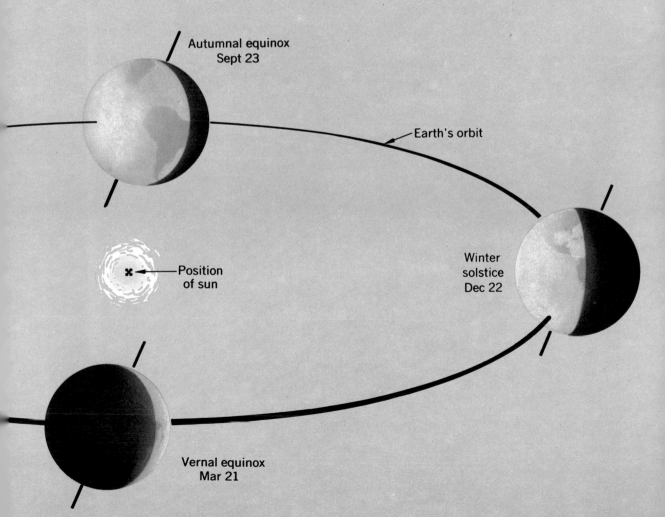

Autumnal equinox
Sept 23

Earth's orbit

Position of sun

Winter solstice
Dec 22

Vernal equinox
Mar 21

The sap of this maple tree has frozen into a sugary icicle in early spring.

regions, the most interesting and easily observed are those that accompany the seasons.

Spring

For many weeks the forest's deciduous trees have stood stark and black against the sky. But one day you notice that a delicate green veil seems to float on their branches: the buds are opening. In the South where there is little or no snow this may happen in late February, but farther north not until May. You feel that spring has come.

Actually signs of the new season may have been evident quite a while earlier. In New England, for instance, "sugar-

Newly formed oak leaves begin to manufacture food as soon as they emerge from their buds. During the spring and summer much of this food accumulates in the trunk and roots and will sustain the tree through the winter.

With the first warm days of spring bloodroots push up
through the litter of an eastern oak forest and unfurl their
graceful petals. The flowers last only a few days, but the
leaves may remain green until late summer.

ing time" begins as soon as the sap rises in the maple trees,
usually while the snow is still deep and the nights are cold.
In many places skunk cabbage appears in swamps and along-
side brooks long before the first tree-buds open, and blood-
root and hepaticas bloom in the woods. Subtle changes of
color herald spring even before it starts: twigs of willows and
beaks of starlings turn yellow in midwinter.

The wild flowers

All winter the forest floor beneath the deciduous trees has
been cold but well lighted. The leafless trees have let plenty
of sunlight shine through to the forest floor. As spring approaches, the sun rises higher and higher, and the sunrays

The delicate hepatica is among
the first of the woodland wild
flowers to bloom in spring.

75

Purple trilliums, spring wild flowers growing in moist
woods, are also called "wake-robins." One flower grows to
a stem and is accompanied by three deeply veined
leaves. Many species of trillium appear throughout the
country in woods with rich soils.

fall more directly onto the forest soil. Temperatures in the
surface of the earth become warmer—in Ohio, for example,
the forest floor reaches its highest daily maximum temperature in April. Then, for a few weeks, before the trees develop
their leafy canopy and cast their shade once more over the
earth, the woodland wild flowers display their brief and gra-
cious bloom.

In the Great Smoky Mountains National Park more than
two hundred kinds of plants flower in April. In the evergreen forests of the West, violets, western skunk cabbage,

sorrel, trillium, huckleberry, and many other plants bloom in early spring. A few days later they are followed by false Solomon's-seal, salmonberry, twayblade, vanillaleaf, and devil's-club.

All these plants are perennials. In early spring they do not have time to manufacture the food needed for sudden growth and flowering; they draw on food accumulated the year before in their underground parts—roots, bulbs, corms, or rhizomes. After their period of flowering, some, such as jack-in-the-pulpit, violet, hepatica, May apple, trillium, and wild ginger, remain green all summer, but others, such as Dutchman's breeches, squirrel corn, trout lily, spring beauty,

The fluttering petals of the Mariposa lily suggest its name, which is Spanish for "butterfly." Native to the western United States, its bulb was once an important food for California Indians.

The rare calypso orchid *(top)* is a spring-blooming, deep-woods wild flower of western and eastern forests. Indian pinks *(bottom)* bloom on wooded western slopes from early April through July.

toothwort, and rue anemone, wither early. In either case the plants remain green long enough to manufacture the supplies of food they will use next year and to produce new buds in which are next year's blossoms.

Chorus in the night

One sign of spring usually is heard, not seen—the chorus of toads and frogs from the woodland's wet places. Rising temperatures in early spring stimulate the first members of the chorus to launch their serenades. Cricket frogs, tree frogs, spring peepers, and wood frogs are among the earliest choristers. As spring progresses and the warm rains of late April drench the swamps and swell the streams, species after species joins the concert, until finally it fades away with the coming of drier weather.

Pleasant though the amphibian chorus may sound on a spring evening, it is not to entertain casual listeners. It serves to promote the courtship of the frogs and toads. Only the male can sing. He produces reverberating sounds by forcing air from his lungs into his mouth and then through tiny openings into a vocal sac that swells out like a balloon from his throat and neck. Each call deflates the sac, and air is drawn into the lungs before the next call.

The calls of the males attract females of the same species, and mating occurs. The eggs are laid in water, except in the case of a few toad species that lay their eggs in moist earth. If the weather is warm, tiny tadpoles emerge from the eggs in two or three weeks and begin the period of growth that will transform the smaller species into frogs and toads be-

TREE FROG

SPRING PEEPER

WOOD FROG

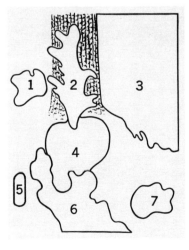

Wild flowers of many kinds, interspersed with other plants, bloom profusely on the floor of an eastern forest in spring:
1. Large-flowered trillium (white)
2. Interrupted fern (green)
3. Wild phlox (blue)
4. Columbine (red)
5. Bishop's cap (white)
6. Foamflower (white)
7. Violet (blue)

79

Swelling its throat until it looks like a bubble about to burst, a male spring peeper—only 1⅛ inch long—calls from the edge of a woodland pond. His peep is so loud that it can be heard nearly a mile away.

fore the end of summer. Larger species, such as the bullfrog, remain in the tadpole stage until the following summer.

About the time the springtime chorus of frogs and toads is beginning, their voiceless relatives, the salamanders, also appear in great numbers, especially in moist forests. By night they scuttle about the forest floor, but by day they remain protected under fallen logs and dead leaves. If you place a few squares of heavy cardboard on the ground in a moist forest area, probably after a few days salamanders will have adopted them as daytime shelters, along with toads, ants, millipedes, and many other small animals that live on the forest floor.

The forest comes to life

Each day the sunlight is warmer, the trees greener, the herbs taller. A greater number of insects appears. Some are immature forms that have just hatched from eggs deposited last fall or summer, others are new adults that have emerged from cocoons or similar shelters; still others are species that have wintered over by staying in crevices or by burrowing into the soil.

Many of the earliest insects, including the wasps and bees that make their appearance on the first warm day, gorge themselves on the pollen of spring-blossoming wild flowers. Soon the air is full of flies, midges, and mosquitoes. Caterpillars hatch and begin feeding on the unfolding leaves of trees and shrubs, while other insects explore the soil and climb the trunks and branches.

Garter snakes come out to sun in open spaces, their mud-stained skins revealing where they have spent the winter. Rattlesnakes and copperheads crawl from their winter quarters deep in rocky crevices where colonies of hundreds may gather during the period of hibernation. Bats fly out of caves and hollow trees at night to feed on flying insects and thus regain the weight they lost during their long winter sleep.

The groundhog (woodchuck) was up and about weeks

Born in spring, gray-fox pups undergo a full four months of extensive training. During this time they learn to climb trees and to hunt for small animals, especially rabbits and mice. The young females usually breed the next season and bear from one to seven pups.

ago—looking for his shadow. At least that is the fable; the fact is that he may emerge during late winter or early spring, look for a mate, and then go back to his burrow for a brief snooze. Now, when the fields and clearings have plenty of green growth, he appears again, this time for several months.

During the winter, bear cubs were born—naked, helpless, no bigger than squirrels. Now they are well developed. They frolic and wrestle on the forest floor, and trail their mother on food-hunting excursions. Litters of rabbits and hares are just now coming into the world, and soon there will be baby squirrels, chipmunks, skunks, raccoons, deer, and all the predators—foxes, weasels, bobcats, mountain lions.

Two tiny white-tailed deer fawns nestle beneath eighteen-inch-high May apple plants. As long as they remain motionless, their dappled coats blend perfectly with the sun-flecked forest floor.

The bird season

Spring is the season for birds. Except for a few species that remain all year, the entire bird community of the forest changes. Each week new arrivals fly in from the South. Some are transients that pause for food and shelter before

Under ideal conditions one female white-footed mouse can produce more than 130 young in a year, and each of these is ready to breed when it is ten weeks old. Were it not for diseases and many predators, white-footed mice might literally overrun the forest.

Opossums are born so small that twenty can fit easily into a teaspoon! They remain in a pouch on their mother's belly for about two months, until they are as large as newborn kittens. Then, strong enough to cling to their mother's fur, they ride piggyback through the forest.

continuing farther north. Others are summer residents that will stay several months.

Most of the early arrivals are males, who travel in advance of the females. Each male selects an area where he can feed and raise his young. This is his "territory," which he defends against intrusion by other males of his own species. If another male comes too close, the defender struts and postures aggressively as a warning sign, ready to attack if the warning goes unheeded.

In addition, the male announces the boundaries of his territory by singing about them, though from the bird's point of view he isn't "singing" at all, but asserting claims and rights. The song also may attract the female when she arrives a few days or a week later.

Some birds do not sing, but attract mates and defend territories by other forms of display. Male ruffed grouse

make a drumming noise by beating the air with their wings, first slowly and then faster, producing a thunder that can be heard at a considerable distance. Woodpeckers hammer on dead limbs or hollow trees.

The battle for nesting space is fought among birds of the same species. Other species are ignored, with the result that the territories of several different birds may overlap. In one tree you find different kinds of birds nesting at different levels. One small part of the forest may contain the nests of a vireo, a thrush, a woodpecker, an owl, and various other species.

What are the advantages of these territorial claims? First, food and space are limited. If great numbers of a given species nested in one area, some might starve because they all would be looking for the same kind of food. Second, and perhaps more important, territorial behavior helps to reduce fighting over food and space, which could interfere with nesting activity. Because birds of different species usually eat different food or search for it in different places, several species can nest in the same territory without competing. Vireos feed in the forest canopy, picking insects from the leaves. Thrushes eat earthworms, caterpillars, and other insects of the forest floor. Woodpeckers dig insects and grubs from the bark and wood of trees. Owls prey on mice, birds, insects, and other small animals.

Among most species of birds, the female is the nest-builder, although the males of a few species help. Once the eggs are laid, incubation begins: the parents, most often the female but sometimes both, keep the eggs warm by covering them with their bodies. When the young birds hatch, the parents work hard to feed them. Indeed, insect-eating birds must bring food to their nestlings every few minutes from dawn to dusk.

You may have heard it said that small birds eat their weight in food every day, but this statement is very misleading and is strictly true only if the nestling is very small and its food very watery (berries, for example). As the young bird grows, it requires more and more food from the parent,

The male ruffed grouse habitually prances on a dead log during mating season. Beating its wings rapidly up and down, it produces a drumming sound that resounds through the forest. When a female appears, the male courts her by ruffing up his neck and tail feathers and strutting back and forth.

Although the screech owl *(top)* might battle another owl for possession of a tree hole to nest in, it does not challenge the woodthrush *(bottom)* that nests in the crotch of a nearby sapling.

but the weight of this food expressed as a percentage of the growing bird's weight goes down from perhaps 50 per cent for a young nestling to about 20 per cent for a bird that has left the nest.

In late spring or early summer, the young birds leave the nest. For a few days the parents continue cramming more and more food into their gaping beaks, but a time comes when the parents cease feeding their young, who must fend for themselves thereafter. Some parents raise a second and a third brood, some do not; the number of broods depends on the species and on the geographic locations of the nesting sites. By midsummer the territorial patterns of behavior have begun to wane. Soon many birds that have been competing fiercely for food and space join together in common flocks. By the end of September most of them have departed on their long migrations to winter homes.

Nesting sites

Some people think that birds nest in the treetops, "close to heaven." But the next time you are in a storm, notice how the high branches of the trees thrash in the wind, and imagine what would happen to an egg laid in a nest on one of those branches. Actually, most birds nest lower down. A study of nesting sites in a forest near Pittsburgh, Pennsylvania, showed that more than one third of all nests were less than two feet aboveground. Half the nests were lower than six feet. Only eight nests out of several hundred were higher than thirty-five feet. One cerulean warbler's nest was at forty-five feet, one blue jay's at fifty feet, and one scarlet tanager's at fifty-four feet, but most nests of even these species were far lower.

In spite of all the birds' care in choosing locations and building nests, however, many eggs and young birds are

A female rose-breasted grosbeak stuffs tidbits of food down the throat of a fledgling.

The prothonotary warbler, named after a papal official who wears a bright orange robe, lives in the swamps of the southern United States. It is the only eastern warbler that nests in tree holes, often in dead willow stumps festooned with Spanish moss, an air plant related to pineapple.

lost in every severe springtime storm.

Another popular belief is that birds prefer the deep forest. Actually the greatest number of individual birds and species nest at the edge of the forest, where field and forest meet. Only a few species, such as woodpeckers, titmice, chickadees, warblers, tanagers, and wood thrushes, live in the forest's interior. The others build their nests in the protection of the tangled vines and shrubs at the forest's edge close to feeding areas in nearby fields.

Spring plant growth

In the warm days and cool nights of early spring tree sap flows freely, carrying water and dissolved food substances upward through the trunks to the buds. Within each bud are tightly folded leaves or flowers—or both—borne on the miniature stems of the new growth. These tiny shoots swell with sap and split the buds apart.

With the coming of spring, food and water in the form of sap flow freely to the buds of such deciduous trees as the sugar maple. Within the buds (a) are the leaves, which now begin to grow rapidly, causing the buds to swell. Then the scales are slowly pushed open, splitting the buds apart, and the leaves emerge (b). One by one the buds open and feathery new leaves develop (c). In a few days the leaves are completely expanded (d).

a   b

The outer covering of the buds is formed of scales. In some species these bud scales fall off as soon as the bud opens, but in others they become enlarged and protect the young stems for several days more. In either case, when they fall the bud scales leave marks around the stem, a ring of scars marking the year's growth. You can usually tell the age of a twig by counting the circles of bud-scale scars on it.

The new stem lengthens; the leaves unfold. This initial growth is brought about not by the creation of new cells but by the enlargement of cells already formed within the bud. As sap flows into them, they expand like balloons being filled with air. Small clusters of cells at the tips of the stems and at places where the leaves join the stems retain the ability to divide and form new cells. These clusters are the beginnings of new buds that will produce next year's growth. By midsummer the tree has completed nearly all its growth for the year. The new stems have reached their full length, and the new buds are completely formed.

c   d

Tree flowers

You are familiar with the beauty of apple blossoms on a spring day. The flowers of cherry and magnolia trees are famous. But other trees have flowers too, although most people never notice them.

The most conspicuous flowers are those pollinated by day-flying insects; this is true of trees and smaller plants. Showy flowers attract insects. In the case of trees, these flowers are usually large and brightly colored, and often fragrant; in addition they most often contain both the pollen-producing or *staminate* parts (often called "male") and the egg-producing or *ovulate* parts (called "female"). Cherry, magnolia, and apple trees are all of this type; so are tulip trees, redbud, black locust, honey locust, sassafras, dogwood, basswood, and various others.

Many trees, however, are pollinated by the wind rather than by insects, and do not have showy blossoms. Usually their flowers are green or brown and are quite inconspicuous.

A galaxy of pollen showers from the conelike blossoms of a jack pine. In spring the air of a coniferous forest may be clouded with billions of pollen grains, but only an infinitesimal fraction of these will fall upon female conelets and thereby produce seeds.

The flowers are either staminate or ovulate, though both kinds may appear on the same tree. The chance that a particular wind-borne grain of pollen from a male flower will ever reach a female flower is very slim, and wind-pollinated trees must produce enormous quantities of pollen to insure reproduction, as some victims of early-season hay fever know very well. Among the wind-pollinated trees are elms, oaks, willows, poplars, hickories, walnuts, birches. Pines, spruces, firs, and other conifers also are wind-pollinated, but they bear cones rather than true flowers.

For a seed to be produced, a pollen grain must reach a female flower of its own species. It must land on the *pistil* of the flower and then develop a *pollen tube* that carries male germ cells down to an *ovule,* or future seed, at the base of the pistil. Once male cells enter an ovule, an embryo plant and food-storing cells develop, the outer tissue of the ovule hardens, and a seed is formed. At the same time the fruit which encases the seed, or often several seeds, develops from the pistil, and sometimes from other parts of the flower, as the petals wither and fall away. Acorns, cherries, apples, walnuts, and many other tree fruits are familiar to everyone, but all broadleaf species bear fruit of some kind. Coniferous seeds are borne in cones, which usually open and release the seeds while they are still on the tree.

Summer

Green is the color of the forest in summer. The early-blooming wild flowers have withered, and by June the leaves of the canopy overhead and the smaller shrubs and trees below have developed fully. The forest is a mass of sun-flecked green foliage.

Now the canopy is the most brightly lighted part of the forest, and also the warmest during the daylight hours. The forest floor, which had received nearly the full intensity of sunlight during early spring, now lies in deep shade. Consequently it is dimly lighted, even on the brightest days, and is cooler by far than the open fields. At night, however, the forest may be warmer than the open countryside, since the reradiation of heat from the earth to the sky is blocked by the canopy.

Summer begins officially on June 21, the longest day of the year (in the Northern Hemisphere), when the sun

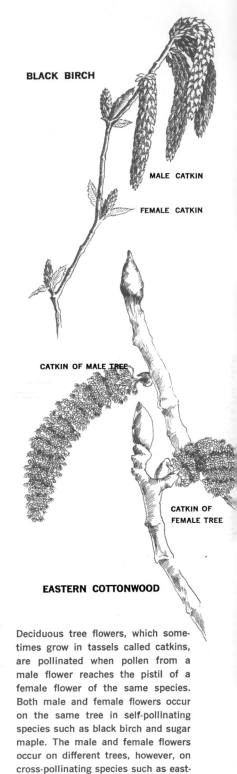

BLACK BIRCH

MALE CATKIN

FEMALE CATKIN

CATKIN OF MALE TREE

CATKIN OF FEMALE TREE

EASTERN COTTONWOOD

Deciduous tree flowers, which sometimes grow in tassels called catkins, are pollinated when pollen from a male flower reaches the pistil of a female flower of the same species. Both male and female flowers occur on the same tree in self-pollinating species such as black birch and sugar maple. The male and female flowers occur on different trees, however, on cross-pollinating species such as eastern cottonwood and American holly.

91

Ponderosa pines rise more than 200 feet above the floor of a western forest. From the soil, water and nutrients travel up the great trunks to the green needles in which food is manufactured. Some of this food flows into the roots and will be used by the tree during the winter when freezing temperatures slow photosynthesis.

reaches the northernmost point of its annual circuit. But by mid-May in many parts of the United States, the fragile colors and melodious birdsongs of spring have given way to the deeper green of mature foliage and the raucous demands of newly hatched nestlings.

The busy plants

Unlike animals, larger plants are rooted in one place all their lives. Yet they are remarkably active.

The most important activity of the forest in summer is photosynthesis. Millions of green leaves use every sunlit hour to produce stores of food at a fantastic rate. Some of the food goes into the flowers, fruits, and seeds of the current year; some is used in the growth of new tissues in trunks, branches, and stems. Some is absorbed by the new buds that will produce next year's flowers. And a good deal flows into the roots and twigs. This accumulated supply will nourish the plant during the long period from autumn to spring when no new food can be manufactured; although the plants are less active in winter than in summer, they still require energy to stay alive.

As a result, the great trees of the forest must produce enormous quantities of food during the summer to meet their own needs. In addition, the plants produce enough to feed the forest's animals, fungi, bacteria, and other microorganisms.

Most plants that grow on land are water-resistant. The exposed parts of trees are covered with bark, and even the leaves are enveloped in waterproof coatings. Yet no plant could live if it were entirely sealed up. The leaf surfaces of trees are pierced by millions of tiny openings, called *stomates*, through which gases pass in and out. During photosynthesis, the plant must absorb large quantities of carbon dioxide from the atmosphere and release large quantities of oxygen. These gases enter and leave the plant by the stomates.

In addition to oxygen and carbon dioxide, water in vapor form also passes through the stomates. The amount of water that escapes through the billions of microscopic holes in the leaves of each tree and through the hundreds of billions of stomates in each acre of woods is staggering. Mountain forests in California may lose a quarter of a million gallons of

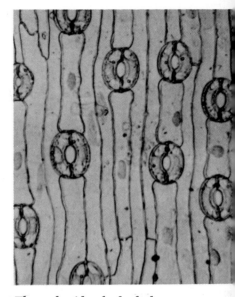

The underside of a leaf, shown highly magnified, is covered with millions of tiny openings called stomates. Cashew-shaped guard cells on each side of the stomate regulate the passage of carbon dioxide, oxygen, and water vapor between the air and the leaf cells.

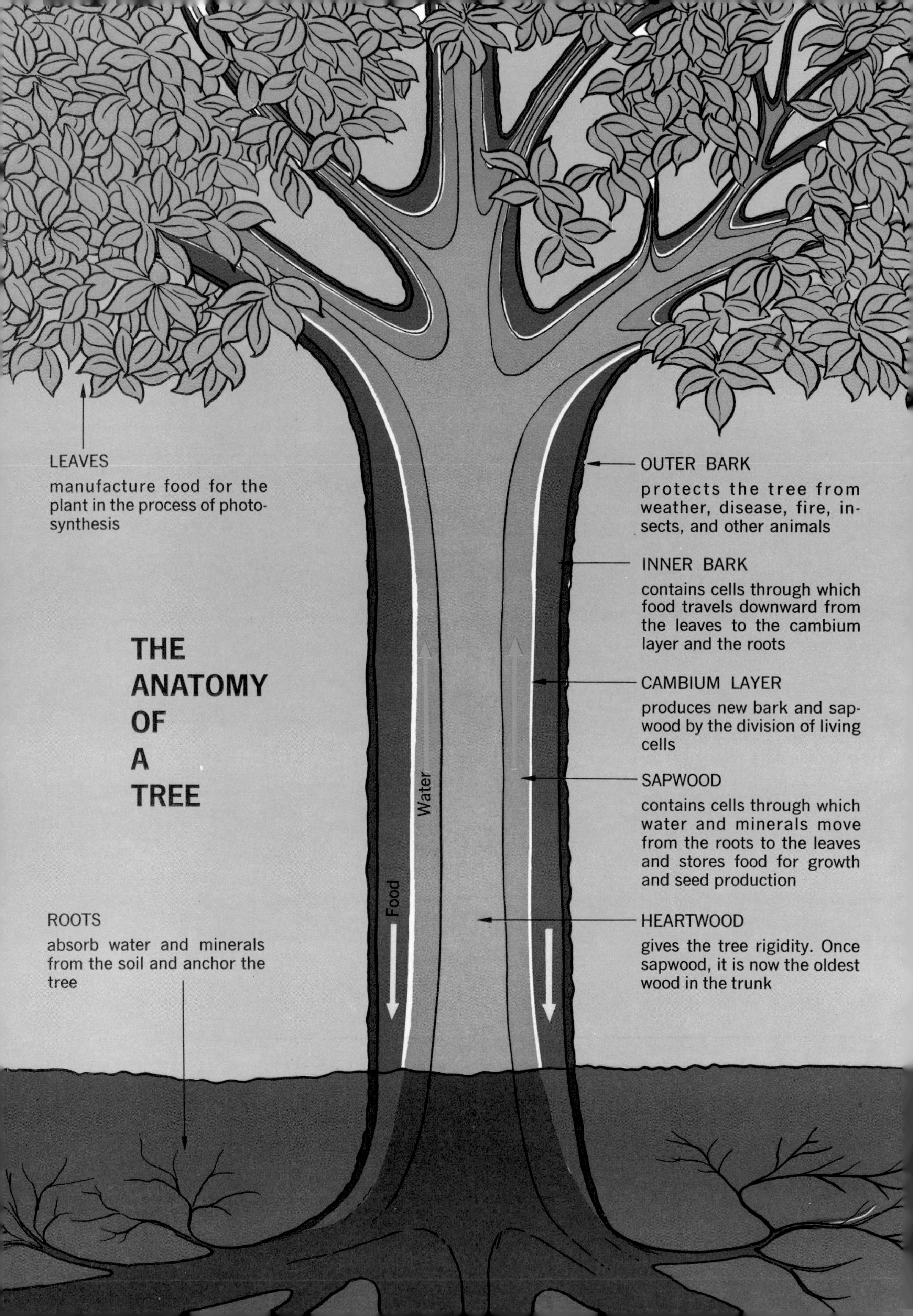

THE ANATOMY OF A TREE

LEAVES
manufacture food for the plant in the process of photosynthesis

ROOTS
absorb water and minerals from the soil and anchor the tree

OUTER BARK
protects the tree from weather, disease, fire, insects, and other animals

INNER BARK
contains cells through which food travels downward from the leaves to the cambium layer and the roots

CAMBIUM LAYER
produces new bark and sapwood by the division of living cells

SAPWOOD
contains cells through which water and minerals move from the roots to the leaves and stores food for growth and seed production

HEARTWOOD
gives the tree rigidity. Once sapwood, it is now the oldest wood in the trunk

Food

Water

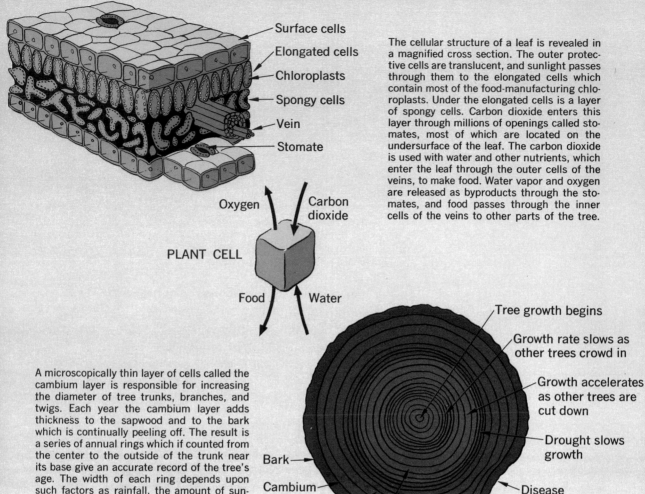

Surface cells
Elongated cells
Chloroplasts
Spongy cells
Vein
Stomate

The cellular structure of a leaf is revealed in a magnified cross section. The outer protective cells are translucent, and sunlight passes through them to the elongated cells which contain most of the food-manufacturing chloroplasts. Under the elongated cells is a layer of spongy cells. Carbon dioxide enters this layer through millions of openings called stomates, most of which are located on the undersurface of the leaf. The carbon dioxide is used with water and other nutrients, which enter the leaf through the outer cells of the veins, to make food. Water vapor and oxygen are released as byproducts through the stomates, and food passes through the inner cells of the veins to other parts of the tree.

Oxygen   Carbon dioxide

PLANT CELL

Food   Water

Tree growth begins

Growth rate slows as other trees crowd in

Growth accelerates as other trees are cut down

Drought slows growth

Disease

Fire

Insect damage

Bark

Cambium layer

Sapwood

Heartwood

A microscopically thin layer of cells called the cambium layer is responsible for increasing the diameter of tree trunks, branches, and twigs. Each year the cambium layer adds thickness to the sapwood and to the bark which is continually peeling off. The result is a series of annual rings which if counted from the center to the outside of the trunk near its base give an accurate record of the tree's age. The width of each ring depends upon such factors as rainfall, the amount of sunlight the tree receives, competition with other trees, and so on. The rings sometimes enclose traces of insect and fire damage, and thus the rings not only indicate the tree's age but also much of its history.

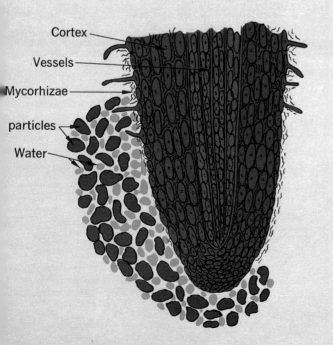

Cortex
Vessels
Mycorhizae
particles
Water

The active root tips of many trees are invaded by a special fungus that passes water and nutrients into the root vessels. Such trees are called **mycorhizal,** which literally means fungus-root. The mycorhizae weave a sheath of threads around the root tip and work their way into the cortex, or root center. This mat of fungus greatly increases the root's water-absorbing surface. In some trees, such as pines, the mycorhizae attack roots only in soils that have poor supplies of nitrogen, phosphorous, and other nutrients. It is believed that the relationship between the fungus and the tree is, in most cases, mutually beneficial and that each plant contributes to the food needs of the other.

Black honey ants make honey from the droplets of sweet liquid that aphids produce when the ants stroke them with their antennae. The honey is then deposited in the bodies of worker ants called *repletes,* which are nothing more than living storage jars, and is eaten when other food is scarce.

water per acre during a year. But the greatest losses occur in the forested stream bottoms of desert areas. During the six warmest months of the year, the trees in one desert canyon lost one and a half million gallons of water per acre!

The process by which water evaporates through the stomates of plants is called *transpiration,* which means "breathing across or through." In a sense, this is a poor name because plants have no lungs and cannot breathe.

Nowadays shortages of water are becoming increasingly troublesome in many areas. Studies are in progress to see how the loss of water through transpiration can be reduced.

Summer insect activity

Insect populations reach their yearly peak during summer. The lush green foliage is an immense food bank for their voracious appetites. By summer's end few of the forest's leaves have escaped at least some damage.

Watch a nearby tree in spring when the leaves come out. At first there may be no sign of insects, but in a few weeks you will begin to notice holes in the leaves. As summer progesses you will find more and more. If you look closely you will see aphids—and perhaps also ants that are tending the aphids for essentially the same reason dairymen tend cows. The aphids secrete a sweet substance that ants eat; the ants tend the aphids, even carrying young aphids unharmed to new sources of food.

Summer is the time to make an insect collection. Caterpillars of all kinds are easy to catch. Some, such as inchworms, occur in such numbers that they may completely defoliate a tree. If you capture and feed a few of them, you will be able to watch their transformation into moths. Perhaps you will find the caterpillars of some of our more striking moths—luna, polyphemus, cecropia—though these species are not so abundant as many smaller ones.

One of the commonest insect pests is the eastern tent caterpillar, whose unsightly nests in the crotches of small

A luna moth rests on a wood fern, its pale-green wings translucent in the afternoon sun. Luna moth caterpillars feed on the leaves of such deciduous trees as hickory, walnut, and persimmon; the adult luna moths, which live only about two weeks, eat nothing at all.

trees are familiar throughout most of the United States. These caterpillars attack not only domestic fruit trees but such wild trees as black cherry and chokecherry, and the damage they do is often serious—too much to be offset by the fact that the caterpillars eventually turn into rather attractive moths. A related species, the forest tent caterpillar, actually does not build tents but clusters in squirming masses on the trunks of the trees it eats, principally broadleaf species such as birch, maple, poplar, and willow.

Butterflies are less common in the forest than in summer meadows, but you are likely to find some species. The tiger swallowtail is common; its larva is a peculiar green caterpillar with two large false eyes on the top of its head. Look for it on the leaves of birch, ash, cherry, or poplar. Another butterfly whose larva feeds on forest trees is the mourning cloak.

Wherever insects are abundant, other insects come to eat them. Spiders, hornets, and mud-dauber wasps are among the busiest of insect predators. They sting other insects and spiders and then take them to their nests; when the young larvae hatch, they have a supply of paralyzed but still-living insect food. Aphids, which are among the most plentiful of insects, serve as food for many predators, including ladybugs and lacewings.

Scores of destructive tent caterpillars swarm over the bark of a wild-cherry tree. These pests acquired their name from the silken tentlike webs they spin in tree crotches to protect their colonies from heat and stormy weather.

Birds in the summer forest

An abundance of insects also means an abundance of birds. Such forest dwellers as warblers, vireos, flycatchers, and tanagers are insect-eaters all year round; but some other species that live in clearings and at the forest's edge and normally eat seeds—particularly the song sparrows, white-throated sparrows, and other finches—may change their eating habits in spring and early summer and hunt insects.

According to one estimate, the birds of New Jersey eat more than 160,000 bushels of insects a day in summer. Generally the young of insect-eating species are hatched at the

The banded garden spider lives at the forest's edge or in cutover areas where grasses and shrubs grow. It feeds on the flying and hopping insects it catches in its complex web. Banded garden spiders, especially young ones, strengthen their silken webs by spinning many filaments across the hubs.

time insect populations are increasing. Born blind and helpless, they may double their weight in the first five days and grow their first coat of feathers by the end of a week.

Many eggs never hatch. Many chicks die in their first weeks of life. Some eggs are infertile or contain imperfect embryos. Sometimes a spell of cold weather kills the embryos, or predators destroy the eggs. From a third to a half of all eggs laid in open nests never hatch. But screech owls, woodpeckers, and other birds that nest in holes lose only about 15 per cent of their eggs.

A young bird that survives the first few weeks of life must still enter its most dangerous period, since 60 to 90 per cent of all songbirds die during their first year. Those that survive the first year have a good chance of living at least one more year, and a few may live much longer if conditions are favorable. Studies show, however, that about half the adult bird population of the forest dies each year.

Birds renew their plumage by molting: the old feathers drop out and are replaced by new ones. During this process, which takes several weeks, the plumage has a ragged appearance, with patches of old and new feathers. Adult birds molt completely after the nesting season. Most male forest birds that are more colorful than the females in the spring and early summer resemble the females during winter after they have molted.

In late summer some songbirds begin to gather in small flocks of perhaps a dozen individuals as a prelude to the fall migration southward. You can find both pure and mixed flocks, though generally the birds of one species tend to flock together. When migration actually begins, however, only a few forest species—such as tree swallows—travel in flocks, while the great majority, including hawks, warblers, thrushes, and vireos, fly southward individually.

Tell-tale feathers identify the plucking stump of a Cooper's hawk. This bird sits quietly on a tree branch until its prey— a smaller bird or mammal— approaches. Then it launches itself into the air, grasps its victim with its talons, and quickly kills it. Next it flies to a stump where it plucks out the bird's feathers before eating it.

Mammals in the summer forest

For the mammals, too, summer is a time for raising young. Fawns born in the spring still wear their spotted coats as

The short-winged, agile goshawk darts through the branches of northern coniferous forests with amazing accuracy. It swoops down on smaller birds with great force and plummets with them to the ground.

101

Upside down on a branch, a red bat sleeps all day. At dusk these flying mammals become active and begin catching insects on the wing.

This black, or cinnamon, bear has come to the edge of a forest glade in Glacier National Park, Montana, to graze on arnicas, yellow dandelionlike flowers. The bear is observing the meadow from the vantage point of a black poplar tree that may serve as a territorial marker. Bears often select particular trees to claw at and bite while emitting growls. The reason for these actions is unclear, but the clawed tree seems to proclaim to the initiated—other bears—that the area is already occupied.

The fisher—a mammal that resembles an elongated fox— lives in boreal and western forests. When it runs at full speed in a bounding gait, it may leave sets of footprints four feet apart.

The long-tailed weasel is most active after dark, when it hunts for small rodents, birds, and cold-blooded animals. The weasel bites a backboned animal at the base of the skull and hangs on until its victim stops struggling.

The bobcat hunts mainly at night for hares and rabbits,
other small mammals, and birds. It also kills fawns
and even, in winter, adult deer. Young bobcats, born
in spring, stay with their mothers until late fall, when
they go off to hunt by themselves.

they browse on tree foliage beside their mothers; their fathers, the bucks, who are solitary except during the mating season, are off by themselves growing new antlers. Families of predators travel together through the forest, chiefly at night. A weasel mother and her young are on the hunt for mice, shrews, young birds. Bear cubs now are well grown but still frolicsome. They accompany their mothers on long rambles through the forest, seeking raspberry patches, ant hills, and the honey-laden hives of wild bees. At night mother skunks and raccoons, their young tagging along behind in Indian file, investigate every nook and cranny of the forest for mice, frogs, snakes, insects, and vegetable tidbits. Young squirrels play tag in the branches of the tree in which they were born, venturing farther and farther from the nest.

As the young animals grow, their first fluffy coats change to the sleeker fur of maturity. The fawns lose their spots, the foxes grow fine bushy tails. Soon most of the young are indistinguishable from their parents, though some species take more than one year to attain full adulthood. They fatten and their coats thicken as the nights gradually become longer and cooler.

Summer has ended.

Young striped skunks, born in early spring, begin to follow their mother on foraging expeditions for mice, insects and beetles when they are about seven weeks old.

Autumn

The wind blows cooler, twilight comes sooner. The sky is a deep lustrous blue, for the sun has moved southward once more and the sunrays strike through the earth's atmosphere at a steeper angle.

When does autumn begin? About September 23, officially —on the autumnal equinox, when the sun crosses the equator. But for most people, autumn begins when the first leaves fall.

Even though there is more sunlight on the forest floor as the leaves of the canopy fall away, the temperature within the forest does not rise. The days are too short now, and the sun gives less heat. With the falling of the leaves, the trees do not absorb as much moisture from the soil, and the forest floor, covered with seeds and dead leaves, remains moist much longer after rains.

This is the season of intense animal activity. There is plenty of food for all. Armies of insects crawl down from the trees and in from nearby fields to hibernate under fallen leaves. Millipedes, earthworms, bacteria, fungi—the decom-

Earthworms rank with bacteria and fungi as important agents in the decomposition of forest litter. In fact, the earthworms in one acre of forest floor may eat their way through and thus mix with the soil as much as ten tons of litter per year.

posers—go to work vigorously, reducing the mass of litter. But as the north wind sharpens and the frost bites more keenly, death overtakes vast numbers, and the tiny bodies of dead insects and spiders mingle with the rubble of fallen leaves.

The acorn

All the seeds and fruits of trees are good sources of nourishment for the wild creatures of the forest, but none is more important than the acorn.

Acorns are the fruit of the oak tree. Although there is some variation among the acorns produced by the fifty-eight native tree oaks in the United States, they all are rich in carbohydrates and proteins. And since, unlike many other nuts, they are soft enough to be eaten by animals that cannot gnaw, acorns are a staple in the diets of many species.

In the mountains of North Carolina, the acorns produced on 2000 acres of oak forest were sampled and weighed; the average yield was estimated to be 129 tons per year. A single oak in Texas bore 100,000 acorns in only one year. But the usual yield per tree for most species of oaks is about 5000 acorns a year.

Years of abundance are called "mast years." Because acorns are one of the most important forest foods, mast years mean a great feast for the forest creatures. Gray squirrels, fox squirrels, raccoons, deer, and black bears feed heavily on acorns. Indeed, acorns provide up to half of their autumn diet. Red squirrels, flying squirrels, white-footed mice, and wood rats are also heavy acorn-feeders. Mice and deer may eat three fourths of all the acorn crop in any given year. In New Jersey weekly counts were made of acorns on the ground in an oak woods during autumn and winter. Deer and mice ate 98 per cent of the acorns within a few days after they had fallen.

Many birds depend on acorns for food. The noted ornithologist W. E. Ritter has written that the California acorn-eating woodpecker "is bound to the oaks of its general area almost as closely as honeybees are bound to flower-producing plants and for much the same reason: their chief staff of life is the fruit—acorns—of these trees." These birds hide acorns in holes they drill in trees—as many as 30,000 acorns in one tree! Other birds that feed on acorns are quail, nuthatches, titmice, starlings, crows, doves, wild turkeys, and pheasants. Altogether, at least 200 species of American birds and mammals are known to feed on oak trees, eating acorns, leaves, twigs, bark, flowers, and buds.

But smaller animals, especially insects, depend just as heavily on the oak. Some insects eat nothing but acorns, living inside until they have consumed the entire inner tissue. Probably the commonest of these insects are the larvae of acorn weevils. Specialists can often tell the past history of an old acorn by the shape of the holes insects have made in it or by other traces of insect life.

In western oak woodlands or coniferous forests containing groves of oak trees, acorn woodpeckers riddle tree trunks with thousands of small holes that they stuff with acorns. Occasionally, however, a woodpecker goes astray and fills the holes with pebbles.

Animal life inside the acorn

Dr. Paul Winston once studied the acorns that fell in a forest near Chicago. He learned that even a small acorn is not a single habitat, but contains a variety of tiny microhabitats, differing in the kind and amount of food they offer as well as in density, moisture, and temperature. And just as there is a plant and animal succession in the forest, there is a succession within the acorn.

A single acorn may be inhabited at one time or another by bacteria, fungi, algae, protozoa, nematodes, spiders, insects, mites, snails, sowbugs, worms, and many other small creatures.

The first of them enter the acorn while it is still attached to the tree. Beetles bore into the acorn; moths and tiny cynipid wasps lay their eggs on the surface. If fungi enter the acorn and soften the tissues, fly larvae can move in. As many as fifty fly larvae may occupy a single acorn.

In late summer or autumn the fully developed acorn drops to the ground. The larvae inside mature and emerge by boring holes through the shell. Scavengers and fungus-eaters enter through these holes and are followed by predators, sometimes whole series of mites, springtails, pseudo-scorpions, and other tiny animals.

All this activity helps decay the interior of the acorn, and it becomes partly hollow or riddled by tunnels. Centipedes, millipedes, earthworms, ants, and other soil-dwellers may enter. At last the softened shell collapses, and the acorn is no more than a small mound of debris on the forest floor. Its remains are worked into the soil by the action of water and tunneling animals.

Why do squirrels store nuts?

Autumn's bounty of fruits and nuts is needed by many birds and animals. Bears and woodchucks feed heavily and build up reserves of fat before they retire to their dens for winter hibernation. Migrating birds also accumulate fatty deposits in their bodies before they undertake the long journey to their winter homes. Chipmunks hide nuts away in their nests; gray squirrels bury them in the ground.

Certainly these activities will help the animals through the coming winter. But do the animals intentionally eat more or gather more to "prepare for winter"?

Because men have always purposely stored up provisions against the time when food would be unavailable, they have ascribed these intentions to animals when they saw the animals doing the same things. But apparently animals have

Using its long snout, the acorn weevil bores into and lays her eggs in young acorns. After the acorns have dropped to the ground, the eggs hatch and the weevil larvae consume more than half their acorns before breaking out as adults.

The abundance of gray squirrels in a forest is dependent on the amount of food available to them. When the nut crop fails in one area gray squirrels migrate by the hundreds to new ranges, sometimes even crossing streams and rivers in their headlong rush.

Migrating tree swallows move in enormous flocks, returning to the same nesting sites year after year at approximately the same time. Tree swallows fly at speeds of thirty or more miles per hour, and their migration flights may take them as far as Central America.

no idea whatever that winter is coming. Their activities are an automatic response to changes in their environment.

Food-gathering and fat accumulation are both apparently responses to the shorter days of autumn. When each day's period of daylight has been reduced to a certain length by the change of season, the animals automatically begin to eat more or to store what they do not eat. Nobody knows exactly how it works, though scientists are studying the problem intently. In general, however, it is safe to say that in the nervous systems of animals some rhythmic process "measures" the hours of daylight, so that an automatic change of behavior is set off when the period of daylight has shortened by a certain amount.

Experiments with flying squirrels have shown that, although they gather nuts and seeds all during the spring and summer, their "take" is only about twenty nuts each night during the first half of the year, but that in autumn it suddenly increases to as much as three hundred nuts a night. The experiments clearly show that this change in food-gathering habits has nothing to do with the change to colder weather as winter approaches; it is triggered by the shortening hours of daylight.

The migrating birds

In the case of migrating birds, science still is unable to answer most of the questions that arise in connection with this fascinating and complex aspect of animal behavior. Many more experiments are needed. However, observations in the field do seem to indicate that many birds begin their southward migration in the fall in response to a change in the period of daylight rather than in response to a change in temperature or a decrease in the supply of readily available food.

Why do birds migrate? How do they navigate on their long journeys, which in some species may extend nearly 10,000 miles? How do young birds born in the North know the route they must follow to reach their southern homes, especially the young birds that travel separately from the adults? Sometimes young birds do not leave the northern nesting grounds until two or three weeks after all the adult birds have flown away, and then they may take routes that differ from those taken by their parents. Certain species of

111

Sugar-maple leaves on a low branch contrast with birch trunks. In autumn when chlorophyll production in the leaf cells declines, the yellow pigments, which have been present all the time, show through, and the leaves turn from green to translucent yellow. When exposed to full sunlight, sugar-maple leaves usually become a brilliant orange.

young sandpipers, for example, travel from Canada to Argentina on a flight lasting several weeks and join their parents in winter areas that they have never seen before. Exactly how they do it continues to remain one of the most puzzling mysteries of nature.

Not all migrations are as long as the sandpipers'. In fact, in the western mountain ranges, songbirds may migrate each spring and autumn only the few miles that take them from the valley to the upper mountain slopes and back. In North Carolina juncos winter in the lowlands and summer on nearby mountaintops. Thus the problems of feeding and nesting that cause some species to migrate thousands of miles have been solved much more easily by others: they find great changes in climate simply by moving up and down the mountainside. On the other hand, many birds, such as woodpeckers, cardinals, and chickadees, do not migrate at all.

Many suggestions have been given as solutions to the mysteries of bird migration, but so far we have few proven answers.

Autumn is the colorful season

Few spectacles in nature can equal the splendor of an eastern hardwood forest in mid-October. Crimson, scarlet, yellow, orange, bronze, purple, brown—the colors flare on hillsides and along roadways with dazzling intensity.

Why? What causes them? Once again science cannot supply a complete answer, but at least part of the answer is known.

We know, for instance, that colors are caused by chemical substances called *pigments,* and that pigments are found in nearly all living things. Several of the pigments that produce the bright colors of autumn are present in leaves from the moment they unfold in spring, but during spring and summer the green pigment in the leaves dominates the other pigments and obscures them. Green is the only color we see as long as the green pigment lasts.

The green pigment in leaves is chlorophyll, the chemical associated with photosynthesis. All summer long chlorophyll is steadily breaking down, but just as steadily the tree replaces it; the leaves of a healthy tree remain green. But when the nights begin to lengthen in September, the production of new chlorophyll diminishes, although the breakdown of old chlorophyll continues. Soon the chlorophyll is gone, and with it goes the green color.

If there were no other pigments in the leaf, it would simply turn pale and colorless when the chlorophyll disappeared. Instead, the bright colors of autumn emerge; that is, the other pigments in the leaf appear as the dominant green pigment vanishes.

Autumn colors, generally speaking, are produced by three pigments, either singly or in combination. The yellow hues come from *carotenoids,* the same pigments that are found in carrots, daffodils, corn kernels, egg yolks, and canaries. The scarlets, lavenders, and purples come from the *anthocyanins,* or "blue-flower" pigments, which are found also in Concord grapes, cranberries, and many other fruits. These anthocyanin pigments are produced anew in autumn leaves, but the yellow pigments were present all summer. Brown pigments called *tannins* give the characteristic glossy brown color to autumn oak leaves. They also are present in tree bark, walnut shells, persimmons, and various bitter-tasting foods.

Weather affects these pigments in various ways. Cool but not freezing nights and warm days favor the production of the colorful anthocyanin pigments, so that Indian summer is

At summer's end, the leaves of red maples produce less and less green coloring matter, chlorophyll. Now the development of red pigments, stimulated by the presence of sugar in the leaf cells, gives the leaves a new glow. The long warm days and cool nights of Indian summer contribute to the kaleidoscope of autumn color.

Autumn in the Green Mountains of Vermont *(following two pages).*

113

often a time when autumn-foliage colors are pronounced. On the other hand, temperatures below freezing kill the leaf cells, so that new supplies of pigment cannot be produced; when this happens, the leaves wither and turn dark. Direct sunlight stimulates the formation of red anthocyanin pigments, sometimes so powerfully that a partly shaded leaf will turn bright red on its sunlit portions but remain green or yellow on its shaded parts.

Other factors, such as the amount of moisture in the soil, also influence autumn colors. From year to year these conditions of weather and soil are different, with the result that no two autumns are alike. Sometimes the fall colors come early, sometimes late, sometimes hardly at all. In the same season trees on opposite sides of a valley may be differently colored, and even on the same tree varying conditions may produce different effects.

But not all the factors that determine autumn colors are external. Some are "inside" the trees; they are inherited factors, a part of the genetic makeup of the different species. Among maples, for instance, the autumn foliage of the red maple is bright scarlet, that of the sugar maple more subdued but still colorful, and that of the black maple a glowing, translucent yellow, while the leaves of the little striped maple are almost as pale as parchment. All other hardwood trees have their characteristic autumn colors, including some, such as elms and butternuts, that wither and turn dark very quickly.

Falling leaves

As the autumn colors begin to fade, the leaves begin to fall. Why? People used to think it was because the leaves were dying, or because temperatures were dropping. But now we know that the length of daylight is a primary factor.

Trees that are kept warm artificially will still lose their leaves as the days shorten. On the other hand, trees lighted artificially for the same period each day will retain their leaves long after the temperatures have dropped to wintry levels.

When the days of autumn shorten, a chemical change takes place in the cells at the base of each leaf stalk, where it joins the twig. The substance binding the cells breaks down, and the leaf eventually falls to the ground.

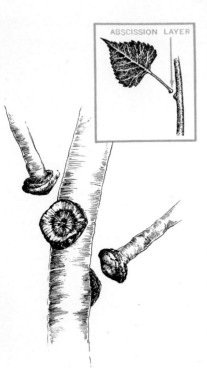

Leaf fall in autumn results from the formation of a separation, or abscission, layer of cells in the leaf's stem. As the growing season ends, the separation layer begins to form, and the shrinkage of the stem, the impact of raindrops, and the formation of ice crystals in the separation layer help break the leaf off.

116

Why is the seasonal behavior of plants and animals more often a response to light than a response to other environmental factors? No one knows for sure, of course, but it is permissible to speculate. Perhaps the reason is simply that most seasonal changes in the environment—changes of temperature, rainfall, food supply, and the like—are quite variable from year to year; the first frost may come early or late, the autumn may be wet or dry. But changes in the hours of daylight are not governed by earthly factors. They are the result of the regular movement of the earth around the sun, and from year to year they do not vary.

Perhaps during their long evolution, the seasonal behavior of plants and animals has become adapted to the most regular feature of their environment, the steady shortening and lengthening of the days.

The end of autumn

Now the winds are sharp and northerly. The forest seems almost empty; most of the colorful birds are gone, though the chickadees call from bare branches. The deciduous trees are bare—no more green leaves now, no yellow or red. Frost-nipped asters and ferns shrivel on the ground, and beneath them lie the unseen bodies of millions of dead insects. Then one day the first wet snowflakes drop from a bleak sky.

Winter has come to the forest.

Near the end of autumn the inevitable first snow falls gently on a Vermont forest, covering the forest floor and heralding the advent of a new season. A tiny beech seedling will be well protected against the oncoming ravages of winter storms by this stand of full-grown conifers.

Winter in the forest is the season of silence. Deep snow has buried the dead leaves and fallen branches and muffles every footfall. There is little birdsong, just the occasional rough call of a jay or a crow or the animated chatter of a passing flock of chickadees, nuthatches, and kinglets. The only other sounds are dry and brittle: a few sere leaves rattling on an oak or aspen, tree limbs creaking in the wind.

Ecologically speaking, winter begins when the last great fall of leaves cascades from the trees and the first hard frost kills the green parts of the perennial wildflowers and ferns. This is the time of death for nearly all insects that do not "winter over" until the following spring; it is the time when hibernating animals are in their deep sleep. In moist areas winter begins when ice coats the ponds and fringes the brooks. In short, winter begins at the point at which there is a marked change from active life to death or dormancy, although many bird and mammal species remain fully active.

In general, winter is a time when energy is expended without being replaced. True, the evergreen trees and smaller plants carry on photosynthesis on days when the temperature rises above freezing; but especially in more northern regions this does not produce enough food to contribute significantly to the total supply. Instead, the life of the forest must subsist on food that was stored up during the green season—in branches, trunks, roots; in seeds, in buds.

The menacing snow

In temperate regions, snow is unique to winter, except on high mountains that may be snow-capped all year.

Deep snows in the western highlands of the Rockies may weigh 5000 tons per acre, or 250 pounds per square foot. This weight alone is a crushing burden for the forest to

Although a heavy snowfall increases the hardships for animals that do not hibernate, it is a protection for those that do sleep away the long winter months. The heavy white blanket acts as an insulator— warmth from the earth is trapped underground and the chill frosts and icy winds are unable to penetrate from above.

Even a light snow makes life harder for animals that remain active throughout the winter. It covers many of the caches they had made earlier and blankets the forest so completely that food is hard to find.

119

bear. Small plants are flattened. Tree limbs are bent or broken, or perhaps bowed down so far that browsing deer and rabbits eat the bark. The blanket of snow is porous enough to allow some air to circulate through it; but when it melts and refreezes, an ice layer that cuts off oxygen from the low plants is formed, and they become easy prey to various types of cold-weather fungi and begin to rot.

The food of larger animals may be covered so completely that it is hard to find. Small seed-eating birds often die, apparently of starvation, since in experiments they have been unharmed by very cold temperatures if they were given enough to eat. Many animals must dig arduously in the snow to find their food. Squirrels hunt for nuts buried weeks earlier. Foxes dig for mice and shrews. Deer paw the snow to find acorns, moss, ferns, and low shrubs.

Snow impedes the movements of animals. Have you tried to run through deep snow? It drags at your feet and trips you. If the snow is crusted, the crust may support wolves or lynx, but not the heavier deer with their small hoofs. At such times deer and caribou are at a serious disadvantage, especially toward spring when they are weakened by a long period of insufficient diet and when the does are heavy with their unborn fawns. Many are now brought down by the predators.

Ground-dwelling species may be trapped under the snow. When the snow melted in Wisconsin one spring, a covey of quail was found dead beneath it, eight birds still in their typical resting circle with their heads toward the perimeter.

Ruffed grouse develop short, stiff, comblike growths along

The red fox *(left)* is slowed down by deep snow and often succeeds in catching a rabbit only when the victim is sick. However, both the Canada lynx and the snowshoe hare *(right)* have broad feet with furry soles, permitting them to move across snow at a rapid rate. Lynxes are so dependent on snowshoe hares for food that when hare populations decline hundreds of lynxes starve.

White-tailed deer trot off single file down a forest trail. In winter, deer create paths in the snow as they travel back and forth between feeding areas. When the snow is deep, they often "yard up" together and browse all the trees and shrubs within reach. Eventually, if cold and deep snow persist, some will starve to death although there may be abundant food less than a mile away.

the sides of their toes in winter, and this helps them to walk on the snow surface. The feet of snowshoe hares are more than twice as large as those of their relatives, the jack rabbits of the Southwest.

Moose, deer, and other large browsing animals often "yard up" in severe weather if the snow is deep. They collect in herds in valley bottoms or small groves of trees. The movements of the herd keep the snow in the yard packed down in narrow trails. In mountain areas such animals frequently move to lower elevations where less snow falls.

When deer gather in yards, they feed on the trees and shrubs around them. The plants may be heavily browsed because the deer that roamed over a broad range in good weather are now concentrated in a small area. If cold and deep snow persist, the deer may remain in their yard until the food is exhausted. The smaller deer, unable to reach the higher branches of the trees, starve first, and the longer the herd is confined to the yard the higher the toll will be. Every winter thousands of deer starve to death.

122

The helpful snow

But the snow may be helpful too. Its protective cover helps plants and small animals withstand the rigors of winter. Many small animals escape freezing winds and sub-zero temperatures by burrowing under the snowcrust. Some birds, such as grouse, dive or tunnel into snowbanks. Mice and shrews escape the cold by tunneling under or through the snow.

Snow is good insulation. The soil in a forest where snow collects deeply may remain unfrozen all winter, while soil in nearby open fields, where the wind blows the snow away, may freeze to a depth of several feet. This may be a matter of life or death to the many animals that spend the winter in the ground. If the ground freezes hibernating species—insects, worms, chipmunks—may die of cold.

The snow blanket also keeps out the noonday heat of the winter sun, with the result that the soil under deep snow is neither as warm nor as cold as exposed soil. The constant above-freezing temperature permits fungi and bacteria to remain active, sometimes all winter, decomposing the litter on the forest floor. Insects and other small animals also may be active. Springtails have even been found more abundantly in winter than in any other season.

Plants that grow in winter

Surprisingly enough, there is a good deal of plant growth in winter. Under the protective cover of snow, the roots of trees and other plants are able to continue elongation. Beneath the litter and in the upper layer of soil, the buds of spring-blooming plants begin to grow early in winter, with tiny leaves and flowers. A few plants even have fully developed pollen in the flowers inside the buds as early as December. Sometimes in more open areas dandelions, chickweed, and other weeds may bloom sporadically throughout the winter.

On trees, the tiny leaves and flowers are folded inside the buds at the ends of twigs and branches. While it might be supposed that the tough scales on the outside of buds protect the embryonic leaves and flowers from freezing, measurements taken with tiny electric thermometers indicate that buds are seldom warmer than the air surrounding them.

In winter, the tough scales of deciduous tree buds protect tiny leaves and flowers which will develop into mature leaves and flowers in spring. During the summer the twig grows longer and a terminal bud, which usually consists of a large flower bud and smaller leaf buds, develops. At the point where last year's terminal bud was, axillary buds sprout above the leaf scars formed when the leaves fall in autumn. The tiny openings, or lenticles, along the twig are pores through which gases move in and out.

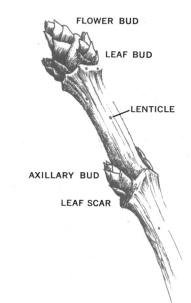

FLOWER BUD

LEAF BUD

LENTICLE

AXILLARY BUD

LEAF SCAR

123

The scales do protect the buds from excessive water loss, however, so that the immature shoots do not dry out and wither in the drying winter winds.

Plants, including trees, continue to absorb some water from the soil all winter. Much more water reaches the soil in winter than in summer; but because the soil is cold or even frozen, the plants do not absorb the water very fast. If the soil is frozen for long periods, some plants may die from lack of water, and this is usually what is meant when people say that plants have been "winterkilled."

The snow cover blocks direct evaporation of water from the soil in winter and therefore is an aid in water conservation. This is one reason that the soil contains so much moisture in spring when there have been long periods of snow cover during the winter. Another is that when the snow melts in late winter and early spring, much of the "meltwater" passes into the soil. In the great transcontinental evergreen forest that stretches across most of Canada and into Alaska, as well as in many mountain forests of the West, water from melting snow may provide moisture during most of the summer. It may be that the survival from ancient times of the giant sequoia trees in very limited areas of the Sierra Nevada is partly a result of the abundant meltwater that runs down the mountains in summer from the snowfields at the top.

In late winter, snow may prolong the dormant period of plants and the hibernating period of animals, especially among the cold-blooded species. Sometimes the snow determines the length of growing season. On high western mountains, for example, the snow reaches depths of fifty feet or more. For weeks after the air temperature rises above freezing each spring, snow still blankets the ground. Thus the short growing season in such high regions is not favorable to the growth of trees, though grasses and wild flowers may abound. Some ecologists believe that the timber line—separating the upper limit of forest growth from the high alpine meadows—may be determined in some areas by the persistence of snow cover in late spring.

Effects of frost on soil

When the ground freezes, a number of things may happen. If the freeze is quick, the frost is loose and granular. But if it is slower, large ice crystals form and water from deeper

Avalanche lilies push up through a drift near timber line in a spruce–fir forest. The snow was blown from the mountain tops by strong winter winds.

levels is drawn upward into the ice layer. Thus water is concentrated near the surface, and when the frost melts in the spring much of the soil's moisture may be lost by evaporation. In an area with little spring rainfall, the soil may be left so dry that tree seedlings and other small plants die. Similarly, many of the minute plants and animals of the soil may perish without water.

The formation of ice crystals causes the surface soil to push upward and then subside when the ice melts. This is called "frost heaving." The movement of the soil may be violent enough to break the roots of tree seedlings. On hillsides frost heave causes a downslope movement of soil that may even cause avalanches in mountainous areas, and in any event leaves the soil loosened and subject to erosion. In one mountain area in California, a third of the rain that fell on a frost-heaved slope in spring ran off over the surface and eroded an average of eight tons of soil from each acre.

In winter the sun is low on the southern horizon much of the time. North-facing slopes are consequently cooler than south-facing slopes. Snow on the northern slopes melts late, and the opening of buds is delayed. Plants on southern slopes, however, develop earlier and often are damaged by late frosts.

Hibernation

As animals have evolved, they have acquired a number of ways of avoiding the hardships of winter. Most birds, a few kinds of insects, and even a few mammals travel to warmer climates where food is abundant. Other animals enter a sleeplike state called *hibernation*.

Most hibernators are cold-blooded animals—protozoa, earthworms, spiders, snails, insects, toads, frogs, turtles, lizards, and snakes. Only a few common warm-blooded animals hibernate, chiefly bears, woodchucks, raccoons, chipmunks, and some bats.

Cold-blooded animals cannot regulate their body heat. As the temperatures of their surroundings fall, their body temperatures fall too. The animals become inactive and enter a state of hibernation in which their body processes slow down drastically.

Snakes and lizards hibernate in rock crevices, in decaying

Black snakes, like other cold-blooded animals, are true hibernators. In the winter they congregate in protected places, such as caves or crevices under rock ledges, where they remain, seemingly lifeless, until the warmth of spring revives them.

125

An eastern chipmunk sleeps soundly in a nest of grasses. Eastern chipmunks are shallow hibernators that sleep intermittently in their burrows during the winter. When awake, they feed on nuts and seeds that they stored away during the summer and fall.

logs, in the burrows of mammals. Turtles that live in ponds submerge into the mud of the bottom, as do frogs and certain salamanders. Hibernating snails dig into the soil and secrete a substance that seals the open end of the shell except for a small hole that allows gases to pass in and out.

Some cold-blooded animals, chiefly insects, have life cycles adapted to the periods of hibernation. The lives of many insects are divided into four stages: first they develop in eggs, next they hatch into *larvae*, then they change into *pupae*, and finally they turn into adult insects. However, both the egg and pupal stages are inactive periods in the life cycle, during which the insect grows but does not eat or move around. Insects that hibernate in cold weather lose less of their yearly feeding time when the egg or pupal state coincides with the period of hibernation. Thus the pupae of many butterflies and moths, hidden in drab *cocoons*, or *chrysalises*, pass the winter inconspicuously among twigs or dead leaves, and the adults emerge only when the weather begins to warm up in spring.

Some colonizing insects such as bumblebees, hornets, and yellow jackets completely die off each fall except for their fertilized queens. The queen hibernates during the winter,

These ladybird beetles, or ladybugs, are hibernating in a
protected spot on a California mountainside. Ladybirds
consume vast quantities of harmful insects in the valleys
during the summer and then congregate on high ground in
the fall. They are covered with deep snow all winter;
but when the snow melts in the spring, they again fly
down to the valleys to feed.

and then in spring lays eggs which start a new colony. Many honeybees, however, live through the cold season. They cluster tightly in their hives, and a few bees at the center of the cluster perform energetic dances that release enough heat to keep the rest alive.

Hibernation reduces the need of some warm-blooded animals for food in winter when food is scarce. Their bodily processes slow down and they require less energy to keep them alive. They can survive for weeks or even months on fat accumulated in summer and autumn.

Mammals are either shallow or deep hibernators. The shallow hibernators—bears, raccoons, skunks, oppossums, badgers—fall into a simple state of drowsiness, from which they may awaken on mild days and leave their dens to search for food. Bears, for example, withdraw into caves or thickets and sleep many hours or days at a time, but their body temperatures remain at about 90°. They are easily disturbed and can wake quickly. Bear cubs are born in midwinter, during the period of hibernation.

Deep hibernators include ground squirrels, woodchucks, bats, and jumping mice. They (not the bats) curl up into a ball with the nose against the belly and the tail coiled over

A black bear rests on its winter bed under a rock ledge in a scooped-out depression lined with leaves and twigs. Bears are shallow hibernators and, if disturbed, will wake up quickly. In their winter dens, females give birth to from one to four cubs in late January or early February.

the head and down the spine. As hibernation begins, the heartbeat slows and the body temperature begins to drop. Then as hibernation deepens, the animal may shiver and move about spasmodically. Finally the eyes close tight and the legs become so rigid that the animal seems frozen.

The body temperature of the deep hibernator may drop sharply, and the heartbeat may slow from as many as 300 per minute, in the case of some small rodents, to as few as three per minute. If the temperature of its hibernating site falls to near freezing, however, the animal's body responds automatically and its breathing, heartbeat, and metabolism speed up, thus keeping its body temperature above freezing without the animal's awakening. But if the temperature of its hibernating site dips below freezing, the animal either wakes up immediately and its temperature rises quickly, or it dies quietly in its deep slumber. It has recently been found that deep hibernators have a special kind of fat tissue called "brown fat" which is important in providing quick energy for the waking-up process.

With the coming of winter the coat of the snowshoe, or varying, hare gradually changes from a brown that blends with the forest floor to a white that makes it nearly indistinguishable from the snow.

Winter diet in the forest

Many animals do not hibernate, of course, and some of them require food that cannot be stored. This is especially the case with the predators. Foxes and wolves are such capable hunters that they can catch other animals crossing open areas of snow-covered ground or they can dig out animals buried beneath the snow. In the North, the fur of some mammals changes from gray or brown to white, thus blending with the snow. This change enables such predators as weasels to stalk their prey with greater ease and such plant-eaters as snowshoe hares to escape predators more effectively.

Moles continue to tunnel underground in winter, but as the ground freezes they cut their runways deeper, below the frostline. Earthworms and some insects also move deeper, furnishing abundant food for the mole. Shrews have a harder time. These little predators must eat almost continuously to stay alive. Often their food supplies are too scanty, and often too the animals freeze to death when temperatures drop suddenly and no snow cover is present to protect them.

The diets of many animals change in winter. Moose, for example, feed during summer on water lilies and other plants in ponds. In winter, when these plants are under ice,

ANIMAL SIGNS
IN THE FOREST

Although many different kinds of animals live in the forests of North America, most are extremely timid, some are active only at night, and some occur in such small numbers in any one location that they are seldom seen. Often we can detect the presence of forest animals only by the tracks they make in snow or moist soil, by their droppings, or by the distinct signs they create when they feed on trees and other plants.

Deer, elk, and moose are browsers. They grasp twigs firmly between their toothless upper jaws and their sharp-toothed lower jaws and snap twigs from branches. Some mammals (among them rabbits, beavers, porcupines, and mice) are gnawers. They have two sharp-edged front teeth in their upper and lower jaws which enable them to make clean cuts without shredding wood. Other mammals, including bears, mountain lions, and bobcats, may not actually eat forest plants, but they damage trees when they sharpen their claws or make territorial signs.

Sapsuckers drill lines of small holes when they feed on tree sap, and woodpeckers chisel out large holes for nesting places. Whatever the sign—browsing or gnawing, scratching or scraping, pecking or clipping—it often can be traced to a particular animal. The tree signs illustrated here are some of the more common ones that might be seen during a walk through a forest.

Deer, elk, and moose leave splintered breaks on woody stems in winter and blunt breaks in spring and summer when the twigs are soft and swollen.

Deer, elk, and moose "ride down" small trees and nibble the tender shoots from their tops. When the trees spring back upright, the eaten areas are higher than these browsers could normally reach.

Meadow mice feed on small branches and stems. They leave small conical cut ends that resemble miniature beaver cuttings.

Porcupines eat large sections of bark from trees. Frequently they attack pine trees, and their large tooth marks are clearly visible.

Ruffed grouse pick off the tender buds of small saplings without damaging the twigs.

Bears tear dead logs and stumps apart to find ants. They leave splintered holes in the log and scatter long fragments of pawed-over trunk.

Browsers also peel bark from tree trunks. The strips of bark, which are pulled upward from the trunk, taper to a narrow point before breaking off.

Rabbits leave on twigs smooth, slanting cuts which resemble knife cuts. Their tooth marks usually appear as narrow grooves where the twig has been clipped.

Beavers gnaw through trunks and large branches. The conical stumps they leave have pronounced channels made by their large incisor teeth.

Pine mice chew irregular patches from root collars and undersides of roots of trees and shrubs. If the mice feed over several winters, the leaves may become discolored and the winter buds smaller than normal.

Bears leave territorial signs on trees by clawing and by slashing with their teeth long vertical gashes on the trunks. Strips of bark heaped at tree bases indicate "bear trees."

Yellow-bellied sapsuckers and their relatives drill lines of small holes around tree trunks and drink the sap that flows out.

For reasons not clearly understood, woodchucks gnaw bark on stems and at the bases of small saplings near their dens and thus produce irregular patchy gnawed areas.

Moose, elk, and deer scrape the velvet from their antlers by rubbing them against tree trunks; the bark of the tree is left hanging in shreds.

Mountain lions sharpen their claws on tree trunks much as house cats do and leave long vertical scratches.

A white-tailed deer stretches its neck to pull down a branch. In areas where winter herds are large, deer may cause permanent damage to trees, shrubs, and seedlings by eating nearly all the branches and twigs within reach.

the moose browse on twigs of trees and shrubs. Many other plant-feeders, such as porcupines, rabbits, and mice, chiefly eat leafy green food or seeds in summer, but in winter nibble the inner bark of trees, often causing considerable damage. Generally speaking, winter foods are sparser and more difficult to obtain than summer foods, and the limited amount of food available in winter may determine the size of the permanent animal population in many regions.

The damage done to vegetation by winter feeders often is quite evident in late winter and early spring. Browsing animals—deer, moose, elk—snap off twigs and even whole branches from trees and shrubs. Sometimes, to reach the tender twigs at the top, they knock down small trees and saplings by walking over them. Gnawing animals such as porcupines frequently strip the bark from the trunks of trees, thus girdling the trees and eventually killing them. Mice too can destroy shrubs by eating bark, as many a rose gardener has learned to his sorrow.

Squirrels also eat bark, and new buds as well, delicately removing the outer scales. Such birds as pine grosbeaks, eve-

This sinuous triple track was left by a shrew that ventured abroad from under the warm cover of the snow below which it digs in search of dormant insects and worms. Shrews have enormous appetites and must seek new burrowing places for food from time to time. Otherwise they tend to stay hidden and only appear on the surface when absolutely necessary, usually at night.

ning grosbeaks, crossbills, and purple finches—all common in northern forests during the winter—also eat the buds of maples, birch, and evergreens. In the West, pocket gophers dig tunnels among the roots of trees, eat the small roots and strip the bark from larger ones. Sometimes they even pull small seedlings downward through the soil into their burrows where they eat the tender stems.

Bears in winter, when they come out of hibernation during warm spells, may strip the bark from trees with their long canine teeth, and they also have the peculiar habit of clawing the bark at "fingertip" height as a sign to other bears that may pass the same way.

The end and the beginning

By now the sun's daily course is noticeably higher in the sky. The weather has turned warmer, at least by day. The sound of running water is heard in the forest: rivulets beneath the snow, icicles dripping from trees and ledges. Alternate thawing and freezing make the snow mushy— what skiers call "corn snow." And perhaps here and there are signs that trees will shortly renew their growth: a smell of pitch in the pine grove, some sap oozing down a sugar maple's bark where it has been tapped by a red squirrel.

For ecologists, this is the end of the old year, the beginning of the new. Soon a faint haze of green will dust the poplars; the first note of the frog chorus will sound from the swamp. The silence of the forest will be broken, and spring —the time of renewal—will be with us once again in the never-ending cycle of the seasons.

In early spring the forest floor slowly thaws, moistening the crisp, dry leaves of the past year, and millions of organisms emerge from the soil and litter to repopulate the higher levels of the forest.

Land of Many Forests

If you were orbiting in an artificial satellite and could look down from several hundred miles above the continent of North America, what would you see? Not much detail, for you would be too high. But assuming the atmosphere were free of clouds, you would notice broad patches and bands of colors defining the outline of the continent.

The darker patches would be the oceans and great inland lakes. Areas of brown or gray in the West would be deserts. To the north would be a region of white—the immense snowfields and ice floes of the Arctic. But a large part of the continent would be green, a mixture of light and dark shades in irregular patterns—the vegetation that spreads across the continent from east to west and north to south.

Suppose your satellite could descend straight down like a helicopter. As you came closer to the earth details would begin to emerge, and you would be able to distinguish between the different areas of vegetation.

You would see that the pattern of forest vegetation resembles the letter **n**. One arm of the **n** extends from Mexico to Canada, with its arch and tail stretching from western Alaska across Canada to the Atlantic Ocean. The eastern

137

arm of the **n** reaches from Canada to the Gulf of Mexico, and between the arms in the open center of the **n** are the broad grasslands of central Canada and the United States.

North America is a region of variety and contrast in both its forested and its unforested areas. Surrounding the network of forests that covers the continent are treeless tundras and grasslands, hot and cool deserts, scrublands and savannas —and each type of landscape has distinctive kinds of plants and animals. In the south there are near-tropics; in the extreme north, frigid polar regions where in some places the ice never melts. There are level stretches of plains, and rugged mountains with peaks often swathed in clouds. There are differences in climate and microclimate—different rhythms of temperature, different patterns of daylength and nightlength, different amounts of rainfall and snowfall.

Land of many plants and animals

Throughout the forested **n** of North America there are thousands of kinds of plants and animals, but no single species is present in every part of the continent. Most species are found only in one particular region, many only in parts of one state, and some only on a few acres of land. No two species of plants or animals are distributed over exactly the same area, and some species never grow together. You would never look for a palm tree in a forest of northern white cedars, nor would you expect to find a polar bear in the company of an armadillo. However, many plant and animal species are distributed in such a way that their ranges overlap considerably. These species form the major associations that are characteristic of extensive regions such as the western coniferous forest and the eastern deciduous forest.

Why aren't most kinds of plants and animals spread uniformly over North America? Largely it is because each species is fitted to certain conditions and cannot survive under others. A plant can grow only where it gets the proper amounts of water and sunlight, where the growing season is the right length, where the soil contains the necessary combination of nutrients, and so on. Yet it does not follow that a particular plant species will always be found where conditions are favorable. Fires, floods, blights, or human beings may eliminate it from a portion of its range. Chance may determine whether a seed or spore reaches an area

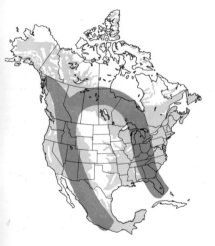

In general outline the forested areas of North America resemble the letter "n." The left leg of the "n" extends from Mexico through the coniferous forests of western United States, Canada, and Alaska; the arch crosses Canada to the Atlantic Ocean; the right leg passes along the east coast through the eastern hardwood forests to the tropical forests of the deep South.

The Kaibab squirrel (*above*) has a dark belly and a white tail and lives only on the Kaibab Plateau—200 square miles of Arizona highland hemmed in by the desert on one side and the Grand Canyon on the other. Across the Canyon, the wide-ranging Abert's squirrel (*below*) differs from the Kaibab; it has a white belly and a gray tail. When an animal is restricted to a small area, such as the Kaibab Plateau, it often becomes distinctly different from other closely related animals.

where climate and soil are favorable. Chance also may determine which species survives in an area suitable for many. If individuals of one or two species become established first, they may prevent the growth of many other species.

If a plant or animal species is to enlarge its range, it must extend through a series of favorable habitats. A barrier to one species may be a migration route for another. High mountain ranges may block the migration of plant species that grow at their bases but serve as migration routes over which arctic or subarctic species extend southward. Oceans are barriers to land plants but migration routes for marine plants and animals. Grasslands may bar tree migration but act as highways for many other kinds of plants.

The forest traveler

Let's tour the forests of the United States. First, however, it will be a good idea to become familiar with the names of the major North American forest regions. They are the western, the northern or boreal, the eastern, and the subtropical. Within each of these broad regions we can recognize smaller divisions. For instance, the western forest region may be divided into the Pacific Coast forest, the Sierra Nevada forest, and the Rocky Mountain forest, and each of these forests has its own subdivisions.

The most common way of classifying a forest is by the kinds of trees that make up the forest canopy. Needleleaf trees form coniferous forests; broadleaf trees form winterbare or deciduous forests; coniferous and deciduous trees growing together form mixed forests. In addition, forests are often identified by their locations. We speak of the northern coniferous forest, the northern mixed forest, the central deciduous forest, and the southeastern mixed forest. Mountain forests may be classified according to the horizontal zones they occupy. The forest at the base and on lower slopes of a mountain is called a *foothill* forest; the forest on the middle slopes, a *montane* forest; and the forest on the upper slopes, a *subalpine* forest. The upper margin of the subalpine forest where the trees become dwarfed is the *timber line*, the boundary between the forest and the treeless *alpine zone* of tundra, bare rock, and snow and ice.

Because many species of trees and other plants need similar conditions of soil and climate, they often grow side by side in a specific region. The names of the trees that are predominant in a forest are used to identify the forest type or *association*. Thus in the eastern United States there are the oak–hickory forest type, the beech–maple forest type, the southern pine forest type, and the hemlock–white-pine–northern hardwood forest type. Over a large part of southern New Jersey pitch pine and black oak form an association as the most prominent forest species. In many places pine is more abundant than oak, and the forest becomes a type called pitch pine–black oak. In other places, oaks outnumber the pines and the forest type is described as black oak–pitch pine.

As we visit the various forests on our tour, we shall use the descriptive terms we have been using all along—we shall continue to speak of broadleaf and coniferous forests, of hardwoods and softwoods. We shall examine the different layers of the forest and notice different associations and successions. And especially we shall try to acquire a "feel" for each forest.

Forests on mountains grow in bands, or zones, the positions and thicknesses of which are determined by environmental factors that vary with altitude. The trees of each zone grow higher on southern slopes where the climate is more mild. Timber line in the central Rocky Mountains begins between 10,000 and 11,000 feet, but northward it decreases some 360 feet for every degree of latitude. In the Canadian Rockies it occurs as low as 6000 feet.

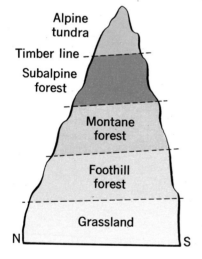

Alpine tundra

Timber line

Subalpine forest

Montane forest

Foothill forest

Grassland

N   S

140

The western forest region

Let's begin by surveying the western forest region. Throughout western North America the forest canopy is composed almost entirely of conifers. This evergreen forest extends up the western arm of the n from the Plateau of Mexico and Baja California to the tail of the n in Alaska, and from sea level to altitudes above 14,000 feet in Mexico. You'll find much of this forest land in the mountainous areas of the Sierra Madres, the Sierra Nevada, the Rockies, the Coast Ranges, the Cascades, the Selkirks and other ranges, and on high plateaus. Above the forests on the higher peaks, such as Pikes Peak and Longs Peak in Colorado, are regions of treeless alpine tundra and perennial snow and ice. Below them, in the dry valleys, are grasslands and deserts.

A major influence on the climate and thus on forest distribution in this region is the Japan Current, a great "river" of mild-temperature water that flows through the Pacific Ocean from Japan to the southern panhandle of Alaska. At the coast this river divides into two streams. One of these streams, the Davies Current, flows northward and warms the southern Alaskan coast so that even in winter temperatures there seldom fall far below freezing. The other stream, the California Current, flows southward and gives the coasts of British Columbia, Washington, Oregon, and California milder climates than regions only a few miles inland.

The winds that blow in from the Pacific Ocean have crossed thousands of miles of open water and are laden with moisture. As they cross colder water near shore and move up the slopes of the coastal mountains, they begin to cool. Water droplets form and produce fogs, rain, or snow on the heavily forested windward mountain slopes and peaks. By the time the air spills over the crests of the higher ranges and flows down the less heavily forested leeward slopes, it has dropped much of its moisture and is becoming warmer.

As the air moves away from the mountains across the lower lands, it retains its warm-dry character, and the lands in this *rain shadow* are typically arid or semiarid—the San Joaquin Valley in California, the Great Basin region, and other large and small valleys between the mountains. In winter these rapidly warming dry winds, known as *chinooks*, can melt a deep cover of snow in a few hours and evaporate much of the resulting water. Grazing and browsing animals such as sheep and deer are probably favored by the removal of

Conifers are abundant in western North America, where only a few kinds of trees make up the canopy of forests. For example, a Rocky Mountain subalpine forest canopy might have only Engelmann spruces and alpine firs, while a Sierra Nevada montane-forest canopy might be composed of sugar pines, ponderosa pines, and white firs.

141

FORESTS OF
THE UNITED STATES
AND CANADA

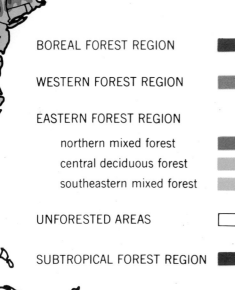

BOREAL FOREST REGION

WESTERN FOREST REGION

EASTERN FOREST REGION

    northern mixed forest

    central deciduous forest

    southeastern mixed forest

UNFORESTED AREAS

SUBTROPICAL FOREST REGION

A GUIDE TO
SOME COMMON
FOREST TREES

On the following two pages are illustrations of eighteen species of trees. Seven of them are abundant in western forests, seven in eastern forests, and four in boreal forests. Information on the range of each tree species and on one or more of its outstanding characteristics accompanies each of the illustrations.

KEY:

WESTERN TREES

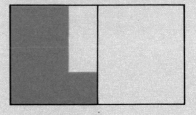

BOREAL TREES

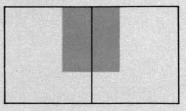

EASTERN TREES

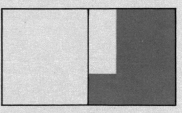

LODGEPOLE PINE
RANGE: Yukon River down Alaskan coast through British Columbia, Washington, Oregon, California, and most of Rocky Mountains.

CHARACTERISTICS: Each scale on the two-inch-long cone has a sharp prickle; cones glossy, light-yellow-brown, often grow in clusters of six or more; bright-yellow-green needles borne in pairs.

DOUGLAS FIR
RANGE: Throughout western mountains from eastern base of the Rockies to Pacific Coast, northern Mexico and mountains of western Texas, southern New Mexico and Arizona north to British Columbia.

CHARACTERISTICS: When young, Douglas firs look like "Christmas trees"; old trees very tall with no branches on the trunks for one third of height; bark deeply fissured and reddish-brown.

WHITE SPRUCE
RANGE: Labrador to Alaska, south to Montana, the Lake states, New York state, and New England.

CHARACTERISTICS: Very short pale-blue or whitish pointed needles that have a foul odor when crushed.

ENGELMANN SPRUCE
RANGE: High Rocky Mountains from British Columbia to Arizona, New Mexico, and the Cascade Mountains of Oregon and Washington.

CHARACTERISTICS: Soft, flexible blue-green needles directed toward tip of twig; in spring dark-purple male flowers and bright-scarlet female flowers look like catkins near the top of the trees.

TREMBLING ASPEN
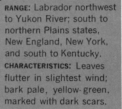

RANGE: Labrador northwest to Yukon River; south to northern Plains states, New England, New York, and south to Kentucky.

CHARACTERISTICS: Leaves flutter in slightest wind; bark pale, yellow-green, marked with dark scars.

TAMARACK
RANGE: Atlantic Coast west to Valley of Yukon River; south from northern limit of tree growth in Alaska and Canada to Lake states, New York state, and New England.

CHARACTERISTICS: Inch-long needles, bright green in summer and yellow in fall, grow in clusters of twelve to twenty.

SITKA SPRUCE
RANGE: Grows along a narrow strip 40 to 50 miles wide and 2000 miles long from Mendocino County, California, to eastern end of Kodiak Island, Alaska.

CHARACTERISTICS: Largest North American spruce; in dense stands trunks of mature trees are free of branches for the first forty to eighty feet; flexible papery cones hang.

SEQUOIA
RANGE: Some seventy groves of five to a thousand trees from North Calaveras Grove, California, east to Lake Tahoe, and south for 260 miles to Deer Creek Grove, California.

CHARACTERISTICS: Trunks of mature trees rise 80 to 225 feet before first limbs; bright deep-green foliage in form of scalelike, sharp-pointed leaves; deep fissures in thick red-brown bark give trunk fluted appearance.

PONDEROSA PINE
RANGE: British Columbia and Black Hills of Dakotas south in Pacific and Rocky Mountain region to Texas, New Mexico, Arizona.

CHARACTERISTICS: Long needles borne in clusters of three; cones grow in bright green or purple clusters; bark on older trees cinnamon brown in papery scales.

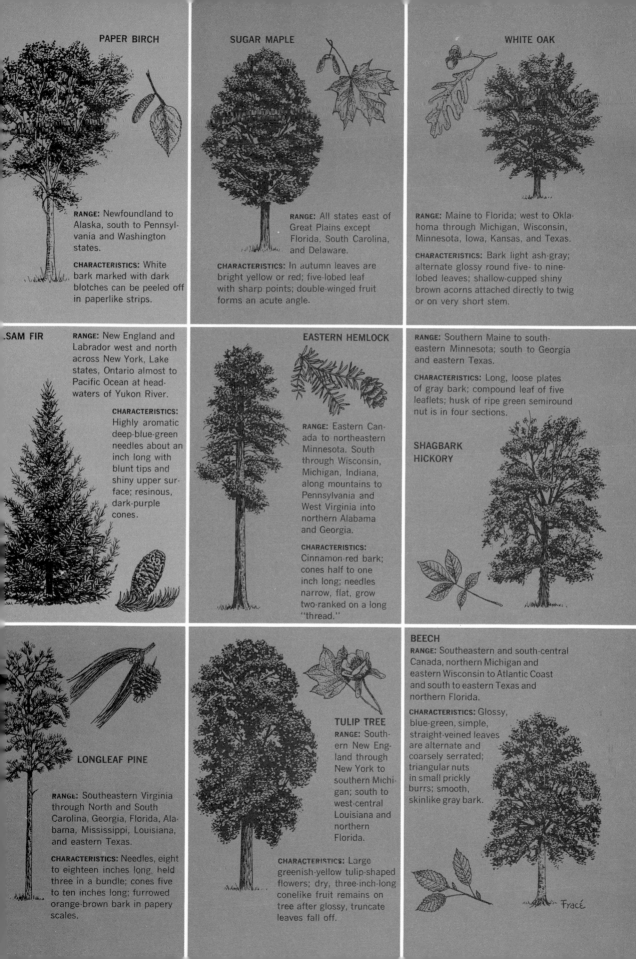

PAPER BIRCH

RANGE: Newfoundland to Alaska, south to Pennsylvania and Washington states.

CHARACTERISTICS: White bark marked with dark blotches can be peeled off in paperlike strips.

SUGAR MAPLE

RANGE: All states east of Great Plains except Florida, South Carolina, and Delaware.

CHARACTERISTICS: In autumn leaves are bright yellow or red; five-lobed leaf with sharp points; double-winged fruit forms an acute angle.

WHITE OAK

RANGE: Maine to Florida; west to Oklahoma through Michigan, Wisconsin, Minnesota, Iowa, Kansas, and Texas.

CHARACTERISTICS: Bark light ash-gray; alternate glossy round five- to nine-lobed leaves; shallow-cupped shiny brown acorns attached directly to twig or on very short stem.

.SAM FIR

RANGE: New England and Labrador west and north across New York, Lake states, Ontario almost to Pacific Ocean at headwaters of Yukon River.

CHARACTERISTICS: Highly aromatic deep-blue-green needles about an inch long with blunt tips and shiny upper surface; resinous, dark-purple cones.

EASTERN HEMLOCK

RANGE: Eastern Canada to northeastern Minnesota. South through Wisconsin, Michigan, Indiana, along mountains to Pennsylvania and West Virginia into northern Alabama and Georgia.

CHARACTERISTICS: Cinnamon-red bark; cones half to one inch long; needles narrow, flat, grow two-ranked on a long "thread."

RANGE: Southern Maine to southeastern Minnesota; south to Georgia and eastern Texas.

CHARACTERISTICS: Long, loose plates of gray bark; compound leaf of five leaflets; husk of ripe green semiround nut is in four sections.

SHAGBARK HICKORY

LONGLEAF PINE

RANGE: Southeastern Virginia through North and South Carolina, Georgia, Florida, Alabama, Mississippi, Louisiana, and eastern Texas.

CHARACTERISTICS: Needles, eight to eighteen inches long, held three in a bundle; cones five to ten inches long; furrowed orange-brown bark in papery scales.

TULIP TREE

RANGE: Southern New England through New York to southern Michigan; south to west-central Louisiana and northern Florida.

CHARACTERISTICS: Large greenish-yellow tulip-shaped flowers; dry, three-inch-long conelike fruit remains on tree after glossy, truncate leaves fall off.

BEECH

RANGE: Southeastern and south-central Canada, northern Michigan and eastern Wisconsin to Atlantic Coast and south to eastern Texas and northern Florida.

CHARACTERISTICS: Glossy, blue-green, simple, straight-veined leaves are alternate and coarsely serrated; triangular nuts in small prickly burrs; smooth, skinlike gray bark.

Fracé

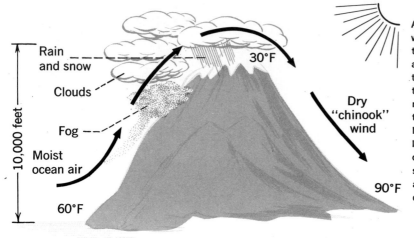

As moist ocean air rises up the windward slopes of western coastal mountains, it cools, water vapor condenses, and clouds form. As the air continues to rise and cool, rain or snow falls on the higher slopes. By the time the air reaches the mountaintops and begins to flow down the leeward slopes, it has lost much of its moisture. Thus the leeward slopes of the mountains receive less rainfall than the windward slopes, and the dry lands to the east are said to be in the "rain shadow" of the mountains.

snow from their food plants in winter, but many small animals in the soil cannot survive the sudden temperature changes.

The western conifer forest is well isolated from cold, dry polar air masses that move down from north-central Canada and from the warm, moist flow from the central portion of the Gulf of Mexico. The Rocky Mountains extend above most of these air masses, so the air is channeled into the continental interior and across the eastern United States. A few polar air masses do spill over the Rockies each winter, but they are modified by the chinook effect and lose their force.

The redwoods

The world's densest and most luxuriant coniferous forest extends along the North American coast of the Pacific Ocean for nearly 3000 miles—from the Santa Cruz Mountains, a few miles south of San Francisco, to Kodiak Island, Alaska. In California and southwestern Oregon, along a strip 500 miles long and 15 miles wide, are found the coast redwoods, the world's tallest trees.

Although bristlecone pines may be the oldest living trees, redwoods are unquestionably the tallest of all living things, some exceeding 350 feet in height. They are resistant to insects that often attack other coniferous trees, and their thick bark protects them from forest fires.

146

Ferns abound in the humid
forests of the Northwest. Two
hundred million years ago these
plants grew in enormous variety
throughout the world. Today
some 6000 kinds are known
to botanists. Only about 250
grow in North America.

"The redwood follows the fogs" noted an observant resident of the redwood forests during the 1800s. The redwood
region does indeed coincide with the zone of frequent summer sea fogs. For several weeks in summer there is almost no
rain along the California coast. The cool air that flows onto
the coast bears too little water and is not chilled enough to
produce rain as it rises over the coastal mountains. Nevertheless, condensation is great enough to produce almost daily
fogs on the windward mountain slopes and in stream valleys.
These summer fogs apparently conserve moisture by reducing evaporation from the forest floor and transpiration from
the trees. In addition, fog droplets collect on the needlelike
leaves of the trees and coalesce into larger droplets. Some of
this moisture may be absorbed into the leaves; some drips
to the forest floor, where it may add to the soil supplies;
some evaporates into the air as the fog dissipates. "Fog drip"
beneath redwoods can be seen easily as you drive through
the coastal mountains. After a few hours of fog, the road surface is wet where trees spread above it but dry between trees.

On mountain slopes, redwood grows in mixture with Douglas fir, tanbark oak, madroño, and several other kinds of trees.
In slope forests the trees usually are spaced widely and seldom exceed 225 feet in height. The dense undergrowth is
composed of California huckleberry, salal, Oregon grape,

thimbleberry, and other shrubs as well as a variety of ferns and wild flowers.

The tallest trees and densest stands of redwood are on level ground at lower elevations, particularly on rich, moist soils of stream terraces in northwestern California. Such lands occupy only about 2 per cent of the redwood region, but these "flat" forests are what one usually thinks of as typical redwood forests. In them the lowest limbs on many trees are 100 feet or more from the ground and the treetops reach 300 feet above the forest floor. The trunks of these towering trees may be 10 to 12 feet thick.

The world's six tallest trees are all redwoods. Think of a tree 350 feet tall—taller than a 25-story building. The tallest known tree was found in 1963 in privately owned Redwood Creek Grove in Humboldt County, California. An expedition of naturalists from the National Park Service and the National Geographic Society surveyed this majestic tree and

In a California forest clearing a bull elk, or wapiti, stands guard over his cows. Female wapitis give birth to one or two calves in May or June; then, herds of calves, cows, and young bulls travel together until breeding season in September and October when the old males claim their cows and run the younger bulls off.

found that the tip of its uppermost twig was 367.8 feet above the soil. The trunk is about 14 feet thick, and has a girth of 44 feet. If the bark were laid flat, it would be nearly as wide as a four-lane highway.

Black-tailed deer are common along forest edges and openings in the redwood belt, and black bears still range the denser, more remote forests. Roosevelt elk are found only in a few places where they have been protected from hunting. Smaller animals are abundant. Runways of shrews and serpentine mounds of soil pushed up by moles are found often near streams. Chipmunks are common where the forest borders on open glades. Chestnut-backed chickadees, brown creepers, winter wrens, and hermit thrushes are a few of the more common birds of the redwood forests.

Muir Woods National Monument

Muir Woods National Monument, named for John Muir, a pioneer naturalist who explored the Sierra on foot, covers less than a square mile and is located only about ten miles north of San Francisco along a beautiful forested canyon where the redwoods grow in their full splendor. The main trail winds among the giant trees interspersed with other big western trees, including Douglas fir. Other trees in the forest are bigleaf maples, California buckeye, tanbark oak, and madroño. The madroño is a beautiful broadleaf evergreen related to eastern rhododendron; its bright orange-colored bark and shiny leaves make it seem, as John Muir once wrote, "like some lost wanderer from the magnolia groves in the South."

Azaleas bloom in the shrub layer of Muir Woods; many western wild flowers grow on the forest floor, among them shooting star, deer-tongue, and salal. You may find tiny yellow violets blooming among ferns at the foot of a redwood whose crown disappears far above the canopy. With luck,

On the floor of coastal redwood forests, shooting stars bloom in early spring. In alpine meadows at altitudes of 10,000 feet, however, related species bloom late in summer after the deep snow has melted.

Shafts of sunlight filter through Muir Woods in California, spotlighting the lush undergrowth of shrubs, wild flowers, and ferns that carpet the forest floor at the bases of giant redwoods. The eerie silence in this ancient forest can cause a visitor to lower his voice to a whisper.

High in the Sierra Nevada, subalpine forests grade upward into snow-covered alpine tundra. A few million years ago the Sierras were raised to their present height by a massive uplift of central California land. Since then the western slopes have been carved out by glaciers and mountain streams, and now steep cliff faces alternate with wooded valleys.

from a section of the tree. Most large trees have fire scars, and many have been hollowed out by fire but remain alive. However, fire is one of the few natural agents that can cause the death of a sequoia—no mature tree is known to have been killed by either insects or disease!

One of the miracles of nature is the growth of a 6000-ton tree from a seed that weighs less than one three-thousandth of an ounce—a 60-billionfold increase. Of course, this gain in weight takes place over many centuries—some of the largest sequoias are 2000 to 3500 years old. Although these trees are ancient by any standard, they are not the oldest living trees known; that title is held by the bristlecone pine. A number of these pines more than 4000 years old were discovered in 1954 on the crest of the White Mountains east of the Sierra Nevada.

Above the foothill and montane zones of the Sierra is the highest forest zone, the subalpine—between 6500 and 11,000

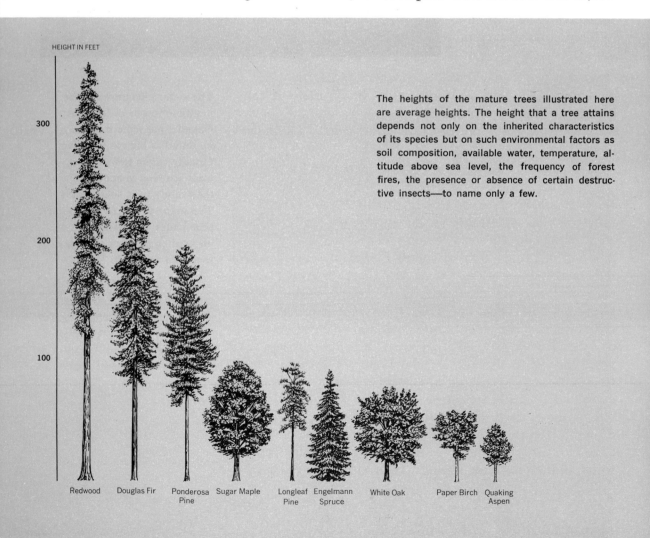

HEIGHT IN FEET

300

200

100

Redwood   Douglas Fir   Ponderosa   Sugar Maple   Longleaf   Engelmann   White Oak   Paper Birch   Quaking
                           Pine   Pine   Spruce   Aspen

The heights of the mature trees illustrated here are average heights. The height that a tree attains depends not only on the inherited characteristics of its species but on such environmental factors as soil composition, available water, temperature, altitude above sea level, the frequency of forest fires, the presence or absence of certain destructive insects—to name only a few.

feet. Red fir is the most prominent tree in the subalpine zone, but in the upper margin are forests of western white pine, lodgepole pine, mountain hemlock, and scattered groves of quaking aspen. Along the upper edge of this zone is the timber line, an irregular, undulating section of vegetation marking the upper limit of tree growth. Timber line varies from about 9500 feet in the northern Sierra to 11,500 feet in the southern. The high peaks above timber line are covered with an open tundra vegetation—scattered low shrubs, luxuriant meadows of grasses and sedges, and growths of cushion-form plants. The vegetation becomes sparser at higher altitudes, and the most elevated peaks always retain some snow and ice, even in summer.

Many large animals move with the seasons up and down through the forest zones of the Sierra. Large-eared mule deer feed in the high-mountain coniferous forests during summer, then move downslope below 4000 feet to the lower

With the accuracy of a weather vane, a stunted Jeffrey pine indicates the direction of prevailing winds at timber line in Yosemite National Park, California. The bark on the windward side of this tree has been polished smooth by ice crystals blown against it.

Camouflaging a carcass, a mountain lion heaps forest litter on a half-eaten deer. It will return in two to four days to feed again and may even stay nearby to guard the cache from intruders.

montane forests and chaparral in autumn. Stalking them in their seasonal movements are mountain lions, each of which must kill a deer at least once a month for food. Black bears, whose thick, shaggy coats range in hue from nearly black through cinnamon-brown to yellow, also move to lower elevations, where oak trees are more abundant, when acorns ripen in autumn.

The bountiful supply of conifer seeds, acorns, hazelnuts, and other tree fruits provides prime food for chipmunks, golden-mantled ground squirrels, gray squirrels, flying squirrels, and chickarees. Preying on these seed-eaters are bobcats and coyotes, who also stalk snowshoe hares, mice, and other small animals. In the montane and subalpine forests you come across other small mammals—yellow-haired porcupines, pocket gophers, weasels, pine martens, bushy-tailed wood rats. Lower in the foothill woodlands are ring-tailed cats, black-tailed jack rabbits, raccoons, gray foxes, and white-footed mice.

Where dying and fallen trees are common you will see several varieties of woodpeckers. The white-headed woodpecker may nest in holes in dead pines or in the thick bark

160

of sequoias and feeds on pine-bark beetles, wood ants, termites, and other insects that tunnel in dying or dead trees. Insects also attack healthy trees in the Sierra forests, killing numerous pines. Particularly on the Sierran east side and in the White Mountains and nearby ranges, Jeffrey pines are stripped and weakened by caterpillars of the Pandora moth. Each spring Indians used to dig broad trenches around the base of the pines to trap caterpillars as they left the trees to burrow in the soil; a single encampment of Indians might gather a ton or more in a few days. The plump larvae were baked or stewed and formed an important part of the Indians' diet.

At home in canyons, caves, and cliffs, the nocturnal ring-tailed cat is rarely seen during daytime. Using its flattened tail as a balancer, this agile acrobat can broad-jump ten feet.

As air from the Pacific Ocean flows up a cool mountain valley, moisture condenses and forms a dense fog which shrouds forests of Sitka spruce, Douglas fir, and other conifers. Rain-bearing winds from the ocean drop as much as twelve feet of water a year along the Pacific-northwest coast.

Olympic National Park

If you travel north from California to Washington state, you will come into the richest growth of the Pacific Coast coniferous forests. Olympic National Park, in the northwestern corner of the state, is a particularly good place to enjoy a temperate rain forest.

The Olympic Peninsula separates the Pacific Ocean from Puget Sound. Virtually surrounded by water, its climate is naturally moist, and its mountains trap the damp seaborne winds and fogs. The average annual precipitation is 142 inches, the highest in the country, and at the peak of Mount Olympus, in the center of the Olympic Range, annual snowfall may be 250 feet.

Most people think of rain forest as a tropical phenomenon, such as the immense growth of the Amazon jungle or Puerto Rico's El Yunque. But the rain forests of the Olympic Peninsula are nearly as lush, although their vegetation is not tropical. They lie on the westward side of the Peninsula, chiefly in sharp valleys scooped out by glaciers.

Douglas fir, western hemlock, and Pacific silver fir are the predominant rain-forest trees; other common trees are western red cedar and Sitka spruce. Occasionally these trees reach 300 feet in height. Their trunks are huge. Beneath them is an understory of small hemlocks, firs, and cedars that may eventually take the place of the abundant Douglas firs, for Douglas-fir seedlings do not grow well beneath their towering parents. But such a succession might require several hundred years and could easily be interrupted by fires or windstorms which might produce ideal conditions for Douglas-fir seedlings.

Bigleaf maples are common in the undergrowth, along with vine maples, salmonberry, salal, Oregon grape, red huckleberry, snowberry, and devil's-club. Many species of moss grow thickly on the tree trunks and on the forest floor, and small club moss (*Oregon selaginella*) hangs like drapery from the tree branches overhead. Mosses, liverworts, and lichens cover many tree trunks. Some lichens are coral-color and tipped with bright red; some are cup-shaped and orange.

The tallest of all spruces, Sitka spruce trees grow rapidly and can reach heights of 200 feet in a century. They grow in a narrow strip along the Pacific Coast from northern California to the eastern end of Kodiak Island in Alaska.

Underfoot are many ferns, including bracken, sword fern, and the spectacular deer fern, and also wild strawberries, bedstraw, wood sorrel, bead-ruby, Kellogg bluegrass, trefoil, foamflower, trailing raspberry, miner's lettuce, wintergreen, vanillaleaf—so many beautiful flowers that the forest seems alive with them. Yet the quiet is deep, and suffused light, greenish or bluish, bathes everything in a dim radiance.

Many fallen trunks, eight feet or more thick, lie covered with moss on the forest floor. These logs rot slowly; some may lie there as long as 400 years. Justice William O. Douglas, who for many years has alternated his duties as an Associate Justice of the Supreme Court of the United States with explorations of the wilderness, has written of the logs:

> They often have tiny rows of Sitka spruce on them, seedlings not more than a few inches high. The seeds that fall on the damp ground, heavy with moss, ferns, and grass, have little chance for survival. Those that land on the old log have head room to grow and lesser competition. This old log will be a

The rain forest of Olympic National Park, Washington.

Appearing to stand on their root tips, Sitka spruces line a log that provided a favorable germination site. As the nurse log decays, the roots of the spruces will fill in the space it now occupies.

nurse to the seedlings for many years. In time they will send their roots down and around the nurse log to the ground. For some years the new trees will appear to be standing on stilts. But in time—perhaps several hundred years later—the nurse log will have decomposed, much of it being absorbed by the new trees. Then the roots will enlarge and fill up the space left by the nurse log. Those that travel the forest on that future day will see giants where I saw seedlings. Not knowing about the old nurse log, they may wonder why it is that these new trees are swollen, distorted, and heavily buttressed at the base. And they may wonder also why the trees stand in a row, giving the effect of a colonnade.

Abundant moisture and moderate temperatures make the dense growth of the rain forest possible. Much of the time, of course, it is raining, though only a fine drizzle comes through the dense canopy, and even after the rain has stopped water may continue to drip from the trees for hours. The warmth of the sea wind nevertheless keeps the dampness from making people uncomfortable, except for a short time in the middle of the winter.

Those who have experienced the quietness and softness of the temperate rain forest are reluctant to leave, but we must get on with our journey.

166

The Rocky Mountain forest

Now we go eastward to another major western range—the Rocky Mountains. Farther from the sea, the Rockies get less moisture than the mountains along the Pacific Coast, and the plant cover is much sparser than in the rain forest.

Nevertheless, the long slopes of the mountains are heavily forested and reach down through the dry country from Canada to Mexico like irregular fingers of green. The Rockies are actually a number of ranges, some almost unconnected with the main system: the Bitterroot Range; the Wind River Range; the Big Horn Range; the Wasatch Range; and, farther south, the Sangre de Cristo and the San Juan Mountains. The higher summits such as Pikes Peak and Longs Peak rise more than 14,000 feet above sea level. As in the Sierra, the larger canyons of the Rockies were filled by glaciers during the Ice Age, or Pleistocene Epoch, and now have sheer, rocky walls. The broad level floors, often covered with meadow vegetation, provide extensive forest-edge habitats for wildlife.

The Rocky Mountains rise from a vast sea of grasses that extends eastward from the mountain front for 500 miles or more. These grasslands are in the long rain shadow of the Rockies. Fossilized plants found in the grasslands suggest that the region was forested before the great mountain chain bulged up and interrupted the flow of moisture-bearing air masses.

On the western flanks, the Great Basin scrub desert, in the rain shadow of the Sierra Nevada and other high ranges, laps onto the foothills. On the lower mountain slopes bushy oaks and low oak trees are mixed with mountain mahogany and other shrubs. Above these broadleaf trees, the forest is composed mostly of conifers.

The foothill zone is piñon–juniper woodland. Several kinds of junipers and nut pines, or piñons, are the predominant trees in these ten-to-thirty-foot-tall "pygmy" conifer forests. These low forests cover the mountain flanks from Mexico to southern Idaho on the west slope and to central Colorado on the east slope. They also spread westward into California on the isolated mountain ranges of the Great Basin. According to one estimate, piñon–juniper forests cover nearly 75 million acres. You will see them in Mesa Verde and Grand Canyon National Parks and in the Funeral and Panamint Mountains of Death Valley National Monument.

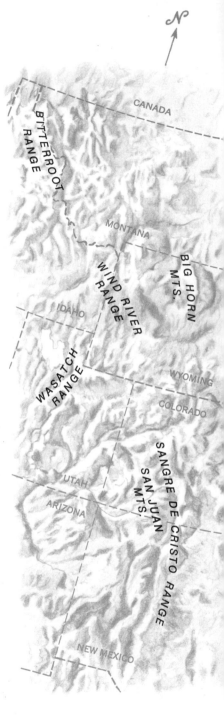

ROCKY MOUNTAIN RANGES

The montane zone begins above the pygmy conifers. Ponderosa pine is one of the characteristic trees in this zone and usually occurs in nearly pure stands in the central and northern Rockies up to elevations of 6000 to 8500 feet. You can find a fine example of such a forest on the Kaibab Plateau, near the North Rim in Grand Canyon National Park. There ponderosa pines form a typical open stand interspersed with aspens and grassy meadows where mule deer graze.

Above the ponderosa pines in the montane zone, Douglas fir is the most prominent tree. On Pikes Peak, the Douglas firs are located between 6500 and 9000 feet; in central Montana they are located between 5500 and 9500 feet. Douglas fir extends along the Rockies from British Columbia and Alberta to Central Mexico. In the southern and central portions of the main Rocky Mountain range, white fir and blue spruce appear with Douglas fir on moister sites. In the mountains of the Southwest, the ponderosa-pine and Douglas-fir forests merge at about 8000 feet and form mixed forests with Mexican white pine and white fir.

In the highest forest zone, the subalpine, spruce and fir

In Grand Teton National Park, Wyoming, alpine sunflowers, scarlet gilias, and wild buckwheat carpet an alpine meadow surrounded by a forest of lodgepole pines, limber pines, and Engelmann spruces.

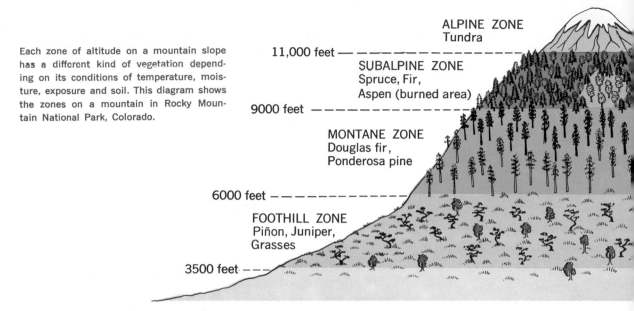

Each zone of altitude on a mountain slope has a different kind of vegetation depending on its conditions of temperature, moisture, exposure and soil. This diagram shows the zones on a mountain in Rocky Mountain National Park, Colorado.

ALPINE ZONE
Tundra

11,000 feet — — — — — — — — — — —

SUBALPINE ZONE
Spruce, Fir,
Aspen (burned area)

9000 feet — — — — — — — — — —

MONTANE ZONE
Douglas fir,
Ponderosa pine

6000 feet — — — — — — — — —

FOOTHILL ZONE
Piñon, Juniper,
Grasses

3500 feet — —

are the principal trees. The spruce–fir forests extend from 9000 to 11,700 feet on Pikes Peak; from 10,000 to 11,500 in northern Colorado; and from 5500 to 7500 feet in northern Idaho and western Washington. Engelmann spruce, subalpine fir, and lodgepole pine are the most abundant trees throughout the main Rocky Mountain region. Characteristic timber-line trees are white-bark pine and subalpine larch in the north, limber pine in the central Rockies, and bristlecone pine in northern Arizona and New Mexico.

Extensive stands of quaking aspen often develop after fires in the ponderosa-pine and Douglas-fir forests of the montane zone and in the spruce–fir forests of the subalpine zone. Aspens sprout from roots when a forest burns and spread rapidly by producing sprouts. Some foresters suggest that several hundred aspen "trees" covering an acre or more may actually be sprouts from a single root system. Usually by the time the aspens have matured, saplings of the predominant trees of the zone have grown up below them and ultimately form a new forest on the site.

Spruce–fir forests blanket the crests of the mountains along great lengths of the Rockies. Relatively few of the ridges and peaks of the United States portion of the Rockies reach elevations high enough to support alpine tundra vegetation. Some of the finest tundra areas, however, are accessible by road in Rocky Mountain National Park, where you can visit several forest zones in an easy day's drive.

169

Yellowstone National Park

With such variety, it isn't easy to choose one place to visit. Rocky Mountain National Park in Colorado might seem a logical starting place, and up on the Canadian border Glacier National Park offers spectacular lake-and-mountain scenery. But the oldest and most famous of the national parks is Yellowstone in Wyoming.

Yellowstone is noted for its thermal displays—hot springs and geysers boiling up from deep volcanic fissures in the ground; no one would want to miss seeing Old Faithful in action. But our present objective is the forest, and within its 3472 square miles Yellowstone contains representatives of several Rocky Mountain forest types. In fact, 90 per cent of the park is forested.

Throughout Yellowstone, particularly in burned-over areas, you will see quaking aspens. At lower elevations—along streams—you will also see other broadleaf trees, including narrowleaf cottonwoods, balsam poplars, and several kinds of willows.

In Yellowstone Douglas firs occur in limited stands below 7000 feet, but the most common conifers are lodgepole pines. In fact, these trees cover nearly 80 per cent of the park, mostly between 7000 and 8500 feet. Lodgepole pines are among the most characteristic trees of the northern and central Rockies. In exposed situations they grow in gnarled or twisted shapes, but in dense stands the trunks are straight and tall—up to 150 feet. Their thick, bushy foliage is bright yellow-green and gives a feathery appearance to the upward-curving twigs. The cones are clay-brown and glossy and may remain on the trees for quite a long time before the seeds are shed.

Fire is a great menace to lodgepoles and other conifers, yet lodgepole forests could not exist without it. Few pine seedlings grow beneath their mature parents, but after a fire the cones of lodgepole pines open and release thousands of seeds on each acre. On the ash-covered soil these seeds germinate, and a new, dense stand of lodgepole pines begins to develop. Without recurring fires, most lodgepole-pine stands would be replaced by forests of spruce, fir, or other trees. Indeed, Engelmann spruce and alpine fir are slowly replacing the lodgepoles in Yellowstone, especially at higher elevations near timber line.

Dense stands of lodgepole pines grow in burned-over forests in higher portions of the central and northern Rocky Mountains.

170

Suppose that you are walking through a lodgepole forest. The pines are tall and straight, rather close together, yet the forest itself is open and unobstructed by low-growing plants. It is dimly lighted, although above the dense canopy the sun is bright. There is almost no noise; the needle-carpeted floor muffles your footsteps. Perhaps you see a flash of color—a beautiful western tanager.

The slope steepens; rocks make walking difficult. You begin to hear a murmuring noise. After fifteen minutes you come to a mountain stream, dashing downward over many boulders. A goshawk wheels away as you approach. You look twice at the stream above you: no, your eyes haven't deceived you. There is a bird walking into the water. It is the water ouzel, also called the dipper, a real clown that can actually swim underwater as it searches for insect larvae and small fish.

You climb to a big rock. The view opens, revealing a distant prospect of mountainsides and meadows. Below you is a stand of aspens; you see the leaves fluttering in the breeze like tiny ripples in a pool and you hear the sound, like a gentle rain falling. High above, a huge bird soars on the air currents; it is one of the few remaining bald eagles. The meadow is bright with Indian paintbrush and blue lupine. Along the stream grow columbine and harebell. If you are lucky, you may see the beautiful blue Rocky Mountain fringed gentian, a species of wild flower which starts bloom-

Water ouzels build their dome-shaped nests of mosses near turbulent mountain streams from Alaska to Guatemala. Ouzels feed on a variety of aquatic insects, often "flying" to depths of ten or twenty feet under water to catch them.

Mates for life, trumpeter swans are the largest of all water fowl and often live as long as a hundred years. These elegant birds, of which only about 600 survive in the western United States, were nearing extinction at the turn of the century. A series of protective measures was taken following 1900, however, and now breeding birds frequently can be seen in Wyoming and Montana.

ing at the beginning of the tourist season in the warm earth of geyser basins and is still blooming in some protected places when the last tourists leave in late September.

If you walk far enough you probably will come to a lake; there are many in the park. And if you are lucky you will find trumpeter swans, for Yellowstone is one of their breeding grounds. Our largest waterfowl, the trumpeter was brought near extinction a few decades ago by hunters; probably at one time there were no more than 150 in the United States. Today, thanks to the protection it receives at such places as Yellowstone, the number has increased to about 600.

In the meadow you see a wapiti, a species whose original range has been greatly reduced. If it is September you may glimpse a huge male with a rack of antlers four feet wide; behind him are his cows, their number depending on his skill in fending off other bulls. In November many of the wapiti migrate to their winter feeding grounds in the National Elk Refuge in Jackson Hole, Wyoming.

The huge antlers of the male wapiti develop from tiny bumps called pedicles. Covered with skin, or velvet, the rapidly growing antlers are about the size of a human hand in April, but may attain a spread of four feet by August. After the breeding season the antlers are shed.

The boreal forest region

The boreal or northern conifer forest forms the arch of the **n** pattern. It extends nearly 4000 miles in a great unbroken band across the continent from the western coast of the Alaska Peninsula on the Bering Sea to the northern tip of Newfoundland and the Strait of Belle Isle on the Atlantic Ocean.

Still farther north lies the windswept expanse of arctic tundra, a land of grasses, lichens, and low, stunted shrubs. Little by little the boreal forest encroaches upon the southern border of the tundra; seedlings spring up among the grasses. Many of them die, and most of those that survive are stunted. Yet very slowly the edge of the boreal forest is extending northward. If, as some scientists believe, our climate is becoming steadily warmer, the trees may cover much of the arctic region tens of thousands of years in the future.

Along its southern margin, the boreal forest merges with the eastern deciduous forest, the central grasslands, and the western conifer forests. In the east it extends southward on

A meandering stream cuts its way through an Alaskan muskeg. Bogs such as this are first overgrown with sphagnum moss that thrives in cold, damp places. The moss, in turn, forms a deep mat on which shrubs and trees can grow.

high elevations in the Appalachian Mountains to Tennessee and North Carolina. In the west it merges with the spruce–fir forests of the Rocky Mountains and the Pacific Coast.

Several times during the past million years the northern reaches of this continent were buried under glaciers. Slowly immense blankets of ice, more than a mile thick in many places, moved southward, planing off hilltops and filling valleys with debris. As a result, the lands now covered by the boreal forest are monotonously flat. The soils are young, their drainage poor. The land surface is dotted with lakes and bogs called *muskegs*. In many places the soil a few feet beneath the surface is permanently frozen, so that the trees are shallowly rooted and easily toppled by the wind.

The major trees of the boreal forest are white spruce, black spruce, tamarack (also known as hackmatack or American larch), and paper birch. These trees are found across North America from the Atlantic coast of Canada to the northwestern coast of Alaska. Some other trees are not so broadly distributed, however, and the boreal forest often is divided into two subregions: the eastern spruce–fir region and the western spruce–pine region.

North of the boreal forests a 2000-mile belt of tundra stretches from Seward Peninsula in Alaska to the eastern end of Pearyland in Greenland. Tundra is much like the land above timber line on mountains. In the southern parts of the tundra belt in protected spots, a few stunted trees grow here and there, but for the most part the tundra is a vast flatland covered with such small plants as reindeer moss, sedges, bilberries, and huckleberries.

175

Dotted with lakes formed during the Ice Age, Superior National Forest in Minnesota merges with Canada's Quetico Provincial Park, Ontario. Southern boreal forests such as this one contain dense growths of white spruces and balsam firs as well as scattered paper birches.

The white spruce–balsam fir region includes eastern and central North America from Newfoundland and Maine to Alberta. Balsam fir often is the most abundant tree, particularly in the southern portions of the region. Usually it occurs on upland sites and is mixed with white spruce. Jack pine is another common tree and often occupies burned-over areas. In the western spruce–pine region, however, balsam fir and jack pine are absent from the forest, and black spruce and larch are less abundant. Here lodgepole pine and subalpine fir appear in the forest; and paper birch, quaking aspen, and largetooth aspen, all present in the East, become much more prominent.

Superior National Forest

To see the boreal forest in its unspoiled majesty we would need to travel into the Canadian northland, along remote streams and trails. It would be a long, difficult journey, so we shall restrict our trip to the United States. Superior National Forest on the northern border of Minnesota will give us a

176

good idea of what the boreal forest is like on its southern edge.

In Superior National Forest the predominant trees are spruces, both white and black, in association with pines and balsam firs. Tamaracks and white cedars are common in muskegs and along streams. We also see such northern hardwood species as paper birch, sugar maple, red maple, yellow birch, and quaking and bigtooth aspen—both called "popples" by the natives.

Chokecherry, elder, and alder are typical shrubs; pipsissewa, twinflower, bunchberry, and pyrola are common wild flowers. Among the less abundant trees are mountain ash, basswood, juneberry, American elm, and hop hornbeam. Among shrubs you may find thimbleberry and bush honeysuckle. Among the dozens of less common wild flowers are wild sarsaparilla and pearly everlasting.

If the climate is warming up, it isn't doing so overnight. In Superior National Forest the growing season is still only a hundred days or less. The snowclad spruces and balsam firs of this kind of forest inspire scenes for Christmas cards; these

The growing season in the boreal forest lasts only about 100 days. Many of the shrubs and herbs on the shady forest floor are green only during this brief period. Scattered broadleaf trees, most of them aspens, also are winterbare.

conifers are well suited to withstand heavy snows and the bitter cold of northern winters. Their resilient branches seldom break under the snow burden.

Our visit is in summer, but try to picture the boreal forest in January. Even at midday the temperature may be close to zero; the air is silent with cold. The sun is so low in the southern sky that forest dwellers may not see it above the treetops for two months or more. The snow is deep; deer gather in herds and trample it down. Foxes wade shoulder-deep through drifts. When a blizzard comes, the temperature may drop swiftly, and the wind may blow with almost hurricane force. Then grouse dive head-first into snowbanks to escape cold, snowshoe hares snuggle deeper into their "forms" beneath the drifts.

In summer the plants and animals of the boreal forest make up for the short growing season by intense activity. Farther north in Canada and Alaska the sun is above the horizon for a long time; when it sets, it sinks only barely below the horizon. The first glimmer of dawn may come at 2:30 in the morning, and twilight lasts until 10:30 P.M. or later. The breeding of northern animals is geared to the short summer. Some, such as minks and weasels, mate in summer and bear their young nearly a year later when the snows have come and gone. Others, such as the snowshoe hares, give birth only a few weeks after mating. Some insects produce many generations during the brief warm season—swarms of black flies and mosquitoes can make campers who forget their insect repellent and mosquito-netting miserable.

Porcupines, skunks, martens, lynxes, even wolves—the north woods have many mammals. Red squirrels are among the commonest, sometimes filling the woods with their noisy scolding. When they gather black spruce cones in autumn, they customarily clip off the twigs just below the treetops, giving the tree a tasseled outline—a tuft of foliage on top, then a bare trunk, then the lower branches. Studies have shown that the bare section acts as a fire barrier. During a forest fire, cones in the upper isolated tufts may escape the

About 30,000 barb-tipped quills cover all but the stomach and face of the porcupine. When in danger, this stocky, slow-moving rodent erects its quills and lashes its tail, driving the sharp spines into anything they touch. Only a few predators, among them the fisher, are skillful enough to kill the porcupine without being injured themselves.

The timber wolf is a beneficial predator that kills only sick deer, moose, and caribou and thus keeps the herds healthy and within normal population levels. The major part of its diet, however, consists of smaller rodents and even insects.

Heads lowered, two bull moose confront each other in a willow thicket *(below)*. Keeping a steady eye on each other they close in *(right)*. Stalemated, both moose stand their ground *(opposite page)*. Confrontations such as this often turn into bloody battles that decide which moose will get an available female.

flames, and their undamaged seeds then fall and begin reforesting the burned section.

Unusual animal noises are associated inextricably with the north woods—loons "laughing" on a lake at night, grouse drumming, a fox yipping, a wolf howling. But undoubtedly the sound that alarms inexperienced woodsmen most is the bellow of one bull moose challenging another to combat.

A large part of Superior National Forest, chiefly in the Boundary Waters Canoe Area, is primitive and uncut. Here spruce and balsam fir tower to heights of nearly 100 feet, astonishing those who have seen only Christmas trees or

ornamental trees used in landscaping. These large trees may be more than 100 years old, with trunks over two feet thick. In the deep forest both spruce and fir lose their lower branches for as much as two thirds of their height, but individual trees growing in the open may keep their conical shape even when they are seventy-five feet or more tall.

The fragrance of the spruce–fir woods, its intense quietness, the outlines of its peaked treetops against a moonlit sky, its beautiful green color in winter, and its unspoiled vastness that still harbors many large animals are good reasons for becoming acquainted with the boreal forest.

The eastern forest region

Now you are in the eastern half of the continent, in New England. Deciduous trees are predominant in the summer-green, winter-bare forest that covers most of the eastern half of the United States and southeastern Canada. Although it is this forest that displays the most brilliant autumn colors, its oaks and maples do not reach the towering heights of the western sequoias and Douglas firs. Mature trees are about 100 feet tall; in some areas the canopy never reaches this height and in others it may be twice as high. The deciduous forest is, however, rich in species, with a great variety of trees, shrubs, and smaller plants.

In New England there are forests in which hemlocks and white pines grow side by side with red maples, sugar maples, birches, beeches, wild cherries, basswoods, elms, and aspens. In boggy places the beautiful soft-needled tamaracks grow. At higher elevations mountain ash is abundant, while on the upper exposed slopes only stunted red spruces and balsam firs can survive the extreme conditions and strong winds. (The strongest wind ever recorded—234 mph—was at the top of Mount Washington in New Hampshire.)

The understory, shrub level, and herb layers in New England forests are often extraordinarily dense. Dogwood, maple-leaf viburnum, high-bush cranberry, blue beech, moosewood, sassafras, dozens of species of ferns, asters, goldenrod, meadow rue, and hundreds of other kinds of plants grow in the New England woods. A mixed forest indeed! And who would want it otherwise—especially in autumn? Then the New England hills are aflame with color: red, purple, orange, yellow, brown, russet—and many other colors; the blue-green of pine and spruce, the rust-green of cedar and tamarack, the rose-tinted white of young paper birches, the silver of beech limbs, and beneath it all the gray-green granite of the land itself.

Because the wilderness was cleared before the beginning

Eastern deciduous forests are composed of a great variety of trees, shrubs, and herbs. Here black maples, their leaves beginning to yellow in the autumn air, frame black cherries, shrubs, grasses, and inconspicuous pine seedlings. In winter the many kinds of deciduous plants shed their leaves, and the pines stand out boldly against the stark landscape.

DOGWOOD

MAPLE-LEAF VIBURNUM

SASSAFRAS

ASTER

GOLDENROD

183

Cutting new material for a dam, a beaver girdles a young tree with its sharp front teeth. These busy mammals do their job so well that they are often used in soil-conservation programs. Two dozen of them built a seven-foot-high dam in Yakima Valley, Washington, which holds 691,500 gallons of water in a pond 90 feet long. Soil carried downstream is trapped by this pond and will eventually form a rich meadow.

of national or state conservation programs, New England offers the visitor few areas of original forest. Nevertheless, the national forests in Maine, New Hampshire, and Vermont, as well as numerous state parks and forests throughout the region, give campers good opportunity for firsthand observation of nature at work reforesting cut-over land. Also, the Appalachian Trail through the White Mountains and the Long Trail through the Green Mountains are excellent routes for woodland hiking, with shelters and camps located at convenient intervals.

Westward and southward extends the hardwood-forest domain. Through New York, Ohio, and into the Midwest as far as Arkansas and Oklahoma, the broadleaf canopy changes as oaks and hickories gradually outnumber maples and beeches. To the south such hardwoods as sycamore, tulip poplar, buckeye, and sweet gum become prominent. Pines, too, are more varied southward: the white pines of New England give way to the pitch pine of New Jersey, the Virginia pine of Maryland and Virginia, the loblolly pine of the Carolinas, the longleaf and slash pine of the southern coastal plain. But in many of these pine woods the understory is oak. If left alone, the oaks would predominate in a few generations, but foresters burn back the oak saplings to keep the valuable pine crop on thousands of acres.

The eastern king snake is one of the most common reptiles in eastern forests. This graceful predator glides over the forest floor, searching among the wild flowers for snakes, frogs, small rodents, and the eggs of ground-nesting birds.

Autumn in the Great Smoky Mountains of Tennessee and North Carolina *(following two pages).*

Great Smoky Mountains National Park

The Appalachian Trail that crosses the peaks of the White Mountains in New Hampshire is a hiker's delight. Running from Maine to Georgia, it threads the entire Appalachian Range, with several subtrails along branching ranges. If you follow it southward, eventually you will come to the region many people call the climax of the Appalachian Highlands: Great Smoky Mountains National Park, which straddles the North Carolina–Tennessee border.

This is the southern forest. Curiously enough it is capped with forests similar to those you saw in the White Mountains. The peaks of the Great Smokies are high enough—about 6000 feet—to have a cool climate. Here the forest is composed of red spruce and Fraser fir, with yellow birch and mountain ash intermixed.

Look out across these mountains from the opening through these red-spruce boughs. It is a view of massive greenery;

A tiger swallowtail butterfly rests on the blossoms of a mountain laurel growing next to a budding rhododendron in the Great Smoky Mountains. Rhododendron, mountain laurel, and other plants form dense growths called laurel slicks that bloom from May through July and turn the forest floor into a carpet of flowers.

The tulip tree or yellow poplar is one of the largest trees in eastern forests, growing to heights of more than 150 feet. Its flowers, as much as eight inches across, develop into conelike fruits composed of winged seeds—a favorite food of cardinals, purple finches, and other birds.

the mountains seem to stretch endlessly, ridges and troughs that have been smoothed by millions of years of erosion. A blue haze rises from the forested billows, giving the mountains their name.

As you descend from the ridge to the lower elevations, you will find more and more southern species, and these are what gives the park its glorious appearance in spring and summer —the flowering trees and shrubs. Among them are magnolia, dogwood, hydrangea, mountain stewartia, laurel, rhododendron, silver bell, serviceberry, wild azaleas in many hues, smilax, sandmyrtle, menziesia, dingleberry, black locust.

In autumn the broadleaf trees, shrubs, and vines blaze with colors that rival the flowers of spring. Maple, oak,

beech, basswood, persimmon, sourwood, Hercules'-club—
their colors range from scarlet to purple to rose to rust.
Sumac and Virginia creeper add their flaming intensity.

Fairly large parts of the park, protected before loggers
reached them, contain the original forest of the southern
highlands and represent the climax growth for this region.

Botanists have listed more than 1300 flowering species in
Great Smoky Mountains National Park, more than 2000 species of fungi, 350 species of mosses and liverworts, and 230
species of lichens. In some areas an observant hiker can make
new discoveries almost every step of the way along a trail.

The trout lily, or dogtooth
violet, blooms in the Great
Smoky Mountains from March
to May. After the flowers
fade, the entire plant dies back
to the ground and new plants
spring up from bulbs the
following spring.

The subtropical forest region

As you continue southward along the Atlantic coast, the
weather becomes warmer, and at last you are in the subtropical forests of North America, at the southern end of

The red fox lives nearly everywhere in North America,
especially in dry upland areas where there are patches of
trees, but also in the tundra, dense forests, swamps, and
even city suburbs. Every nine or ten years the numbers
of red foxes increase, partly because of an expanding
rodent population. When this happens rabies outbreaks
often kill thousands of red foxes.

191

In the swampy coastal plains of the
southern United States, especially in
Florida, are extensive forests of bald
cypresses intermixed with tupelos and other
plants adapted for growing in wet soil and
water. Mature bald cypresses often are 75
or 100 feet high and are nearly unique
among coniferous trees because they shed
their foliage during the winter. When the
needles drop off in clusters attached to
branchlets, the entire tree has a "bald"
appearance relieved only by masses of
Spanish moss hanging from its branches.
The roots of the bald cypress often emerge
from wet soil or shallow water as conical
projections called *knees* and may stand
several feet high about the base of the
trunk *(right)*. There is some doubt as to the
function of these bark-covered knees. They
may help supply air to the water-logged
root system, or they may simply help
anchor the tree in the wet soil.
Cypress forests support a varied
population of wild creatures. Among these
the anhinga *(left)* is one of the unusual
birds. An expert swimmer, it paddles
completely submerged except for its
snakelike neck through the swamp waters.
Herons, egrets, and wood storks are also
abundant. American alligators cruise
through the water or rest on logs or high
ground, sunning themselves. Foxes, bobcats,
raccoons, otters, deer, and other mammals
live in the dense undergrowth of ferns,
hibiscus, water lettuce, and other plants.
Largely as a result of the efforts of the
National Audubon Society, some 6000
acres of cypress forest called Corkscrew
Swamp have been set aside in Collier
County Florida, as a wildlife sanctuary.
Visitors to Corkscrew Swamp can see the
interesting wildlife of a cypress forest by
walking along extensive boardwalks.

In a sea of sawgrass, hammocks of palms form green islands in the Florida Everglades. In spring thousands of migrating birds fill the trees.

the eastern arm of the n. Winter never really comes to these forests; temperatures in the Florida Keys region, a chain of islands that extends into the Gulf of Mexico from the tip of the Florida peninsula, are warm throughout the year, with only slight seasonal differences. The main seasonal change is in rainfall—three to four times as much rain falls during the wet season (May to October) as during the dry period from November to April.

The transition from temperate to tropical vegetation is gradual. Some southern species range northward into central Florida and even farther, and many temperate plants flourish out on the Keys. Southwest from Miami, nearly 200,000 acres of limestone lands between the southern Everglades and coast marsh prairies were covered by widely spaced slash pines and a sparse growth of low shrubs, but now much of the forest has been destroyed by residential construction and farming. Where the forest survives, many tropical trees and shrubs grow with the pines. Similar slash-pine forests are found on many of the Keys.

194

Everglades National Park

Everglades National Park in Florida covers 2100 square miles of land and water, the third largest national park in the United States (only Mount McKinley National Park in Alaska and Yellowstone National Park in Wyoming are larger). There are many campsites, boating facilities, nature trails, and other aids to visitors—including museums and guided tours.

The Everglades is a vast marshy area in which the most prevalent plant is Jamaica sawgrass, a type of sedge with sharp, fine-tooth edges. The land in the park is still much as it has been for centuries, but conditions are threatened because the water table is slowly dropping. Much of the adjacent land has been drained by canals and is now cultivated.

Two kinds of tropical forest are found in the Everglades—*hammock* and *mangrove*. Hammocks are dense forests of broadleaf trees, often mixed with palms, each stand covering only five to ten acres. When they are surrounded by marshes or low pine forests, the tops of the hammocks appear as islands above a sea of lower vegetation. These hammocks

Thriving in salty sand, the red mangrove, called the land-building tree, traps seaweeds and other ocean debris in its tangled roots. Eventually a sandbar with a single mangrove growing on it might become a small island.

are formed by fifty or more species of tall trees, small trees, and shrubs, many of them evergreen. As many as 40,000 trees and shrubs may grow on one acre of hammock on the Keys; shrubs and small trees form a junglelike undergrowth so dense that it may be impossible for you to see farther than ten or fifteen feet in any direction, even upward.

The thick, nearly impenetrable growth of the hammock forest casts a deep shade, almost entirely checks the wind, and produces a moist microclimate. These conditions are ideal for the luxuriant growth of air plants, vines, and ferns which make hammocks even more junglelike. Growing with them are exotic trees and shrubs—strangler fig, gumbo limbo, wild tamarind, poisonwood, red ironwood, and the magnificent royal palm. On the mainland these tropical species are mixed with temperate live oaks, mulberries, hackberries, and red maples. Adding to the lush quality of this forest are the small trees—lancewood, papaya, paradise, and fish poison—and many rare and showy orchids.

Mangroves grow in the salty waters of the Atlantic Ocean and the Gulf of Mexico; they are the northward extension of a forest that lines coasts of tropical lands around the world. The dense forests they form withstand constant buffeting by waves and often by hurricanes, and they are living breakwaters that protect the shores from wave erosion. They also aid in building new lands farther out into the sea by gradually collecting mud, peat, and shells around their many roots.

Most of the mangroves grow as small trees or spreading shrubs, but some are more than two feet thick and over sixty feet tall. Only three of the many kinds of mangroves grow in Florida. Red mangroves form the outer zone of mangrove swamps and may extend far offshore in the shallow waters around the Keys. They develop an extensive tangle of prop roots that resemble curved rods extending out from the trunk into the sea bottom. Another peculiarity of the red mangrove is that its seedlings germinate while the elongated fruits still dangle from the parent tree. Long roots sprout and can anchor the new seedling rapidly once it falls and floats to a suitable growing site.

Black mangroves form a middle zone, growing between the tide marks, so that they are exposed part of the day and stand in salt water at high tide. The black mangrove lacks prop roots, but many vertical extensions that resemble asparagus tips grow up into the air. White mangroves are less

The largest land snail in North America, the two-inch-long *Liguus* makes its home in hammock forests. The colors of its beautiful shell vary from hammock to hammock. Some misguided shell hunters have combed hammocks for these snails and then have burned the trees to kill any remaining. In this way they have added one-of-a-kind shells to their collections.

196

common. Their most distinctive feature is thick oval leaves that look the same on both sides.

It is impossible to do justice to this exotic wilderness in a brief description. Strange birds drift across the sunsets or stalk the marshland: egrets, ibises, limpkins, spoonbills. Pelicans dive for fish in the bay; purple gallinules walk on lily pads. Alligators sun themselves on grassy banks. Frigate birds, gulls, and terns glide effortlessly on the winds. The

The zebra butterfly lives in the dense junglelike growth of southern hammock forests. The caterpillar of this insect feeds on the leaves of passion flowers. The bold black and yellow stripes of the adult warn would-be predators that it is unpalatable.

indigo snake, the longest in the United States, winds like a ribbon of satin among the ferns.

But one of the animals most beloved by visitors to southern Florida is the rare key deer found only on the central Keys. It is a tiny subspecies of the white-tailed deer that grows only about two feet high at the shoulder and weighs about sixty pounds. Who could resist an animal with all the deer's elegance and charm, and hoofs not much bigger than a postage stamp?

The end of our journey

In this brief tour of North American forests we have by-passed many interesting kinds of forests and many areas of spectacular beauty—the oak–hickory forests of the Ozarks, small forests of Monterey pine along the southern Pacific Coast, aspen parklands in southern Alberta and Manitoba, among others. As you journey out into the forests of North America, you will come to know better those you merely glimpsed in this book. But even if your travels are limited to places near your home, you will find that every visit to a forest can be a new adventure. Each forest is unique and each changes from hour to hour, from day to day, from season to season, and from year to year. The forest is the scene of continual battle, with quick and often violent death awaiting the unwary or poorly defended animal. Trees and other plants also compete in this eternal struggle, but their conflicts usually are subtle, often are waged beneath the soil, and may require months or years to complete.

Against the trunk of a tree that may tower more than 300 feet in the air a young redwood seedling begins its journey upward from the forest floor.

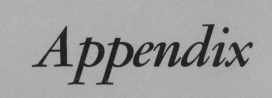

The National Forest System

The 154 National Forests scattered across the United States are areas of many uses. Applying sound conservation principles, the United States Forest Service manages National Forests with the goal of yielding as many public benefits as possible, both present and future. Primary purposes of National Forests are the production of wood and protection of water resources. Although a great deal of timber is cut each year in National Forests, the harvest is never allowed to exceed the annual growth increase. Livestock are also permitted to graze on National Forest lands, though not so many as to endanger continuing productivity of the land. The Forest Service also improves habitat to make it more attractive to wildlife.

For the benefit of the millions of people who visit National Forests each year, outdoor recreation is another major present-day use of National Forests. Picnic areas and campsites are maintained, as are facilities for swimming, boating, and winter sports. Hunting and fishing are encouraged. Exhibits, nature trails, and interpretive programs at many of the forests help visitors understand forest management and ecology.

In spite of extensive use of National Forests for timber production, millions of acres are still in relatively natural condition. Especially in the more remote sections set aside for preservation as "Wilderness Areas," the National Forests offer splendid opportunities for observing and enjoying forests and all the creatures that live in them.

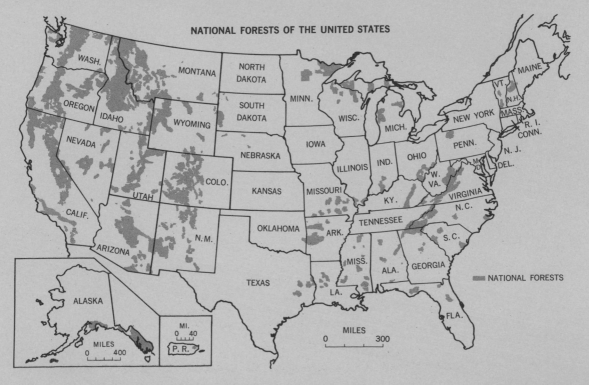

NATIONAL FORESTS OF THE UNITED STATES

Outstanding Forests of the National Park System

Although National Parks and National Monuments occupy less than 1 per cent of the land area of the United States, they represent the best of the American landscape. Areas of unique scenic and scientific value, they have been set aside for the enjoyment and inspiration of present and future generations. Forests are major attractions of many parks; indeed, some of the finest forests in the country are found in the National Park System.

The National Park Service, official custodian of the parks and monuments, provides a variety of facilities for the pleasure and convenience of park visitors. Roads, lodges, and camping areas are maintained at most of the areas. Hiking trails within a single park often total hundreds of miles. Museum exhibits, marked nature trails, and naturalist programs help interpret the natural history of the parks. But the primary duty of the National Park Service is to maintain the parks in their natural condition. All National Parks and National Monuments are treated as sanctuaries in which nature operates with a minimum of human interference. As a result, the National Parks and National Monuments provide a series of living laboratories where the relations between plants and animals and their environment can be observed under ideal conditions.

Areas of the National Park System with forests of special interest are described in this section.

Acadia National Park (Maine)
An island park with an unusual combination of sea, rocky cliffs, mountains, and lakes. Cadillac Mountain (1530 feet) is the highest point on the Atlantic coast. Forests are northeastern spruce–fir; northern hardwood; or a mixture of the two. Forest fire burned thousands of acres in 1947; area now covered by early successional stage of shrubs and saplings, providing abundant food for white-tailed deer. Watch for cormorants, guillemots, and other sea birds.

Big Bend National Park (Texas)
Rugged desert and mountain wilderness on the great bend of the Rio Grande; river passes through three impressive canyons. Besides desert vegetation in hot lowlands, unusual conifer forests on cooler uplands of the Chisos Mountains; mixed with Douglas

COLLARED PECCARY

fir and ponderosa pine (native to country far to the north) are Arizona cypress, piñon pine, and other trees typical of the Southwest. The piglike collared peccary (*javelina*) can be seen at Big Bend.

Blue Ridge Parkway (Virginia, North Carolina)
A scenic parkway almost 500 miles long, on crest of the Blue Ridge Mountains; when completed will connect Great Smoky Mountains and Shenandoah National Parks. Northern hardwood and Appalachian oak forests; spruce–fir on crests, including two fine stands of Fraser fir. Watch for remains of blighted chestnut trees; forest succession on abandoned fields. (Another scenic highway in the National Park System, the Natchez Trace Parkway, eventually will link Nashville, Tennessee, and Natchez, Mississippi.)

OPOSSUM

Chiricahua National Monument (Arizona)
Mountain wilderness of grotesquely eroded rock formations. Rising like a green island in arid grasslands, the mountains harbor many plants and animals. South-facing slopes are desertlike; cooler north slopes and shaded canyon bottoms are covered with dense vegetation. Oak–juniper woodland; some typical southwestern conifers; and several distinctive Mexican species. Chiricahua Apaches once lived in the area; their most famous warrior was Geronimo.

Glacier National Park (Montana)
Superb Rocky Mountain scenery, including 60 glaciers and 200 lakes. Lodgepole pine is the most common tree, mainly at lower elevations. Alpine fir and Engelmann spruce predominate at higher elevations. Significant differences between east and west slopes. Forests on east slope are typical Rocky Mountain types; those west of the Continental Divide resemble forests of the Pacific Northwest. Western hemlock, western red cedar, and western white pine, found here, do not occur in other Rocky Mountain parks. Look for mountain goats on rocky crags above timber line.

Grand Canyon National Park (Arizona)
Canyon of the Colorado River, four to eighteen miles wide and a mile deep; the world's most spectacular example of erosion. Striking changes in vegetation from bottom to top. Canyon floor is warm, desertlike, with cottonwoods and willows along streams. On slopes within canyon and on South Rim is an extensive forest of piñon pine and juniper. Some ponderosa pine on South Rim; Douglas fir grows on shaded steep slopes just below the canyon rim. North Rim is 1000 to 1500 feet higher than South Rim and has a magnificent open forest of ponderosa pine intermingled

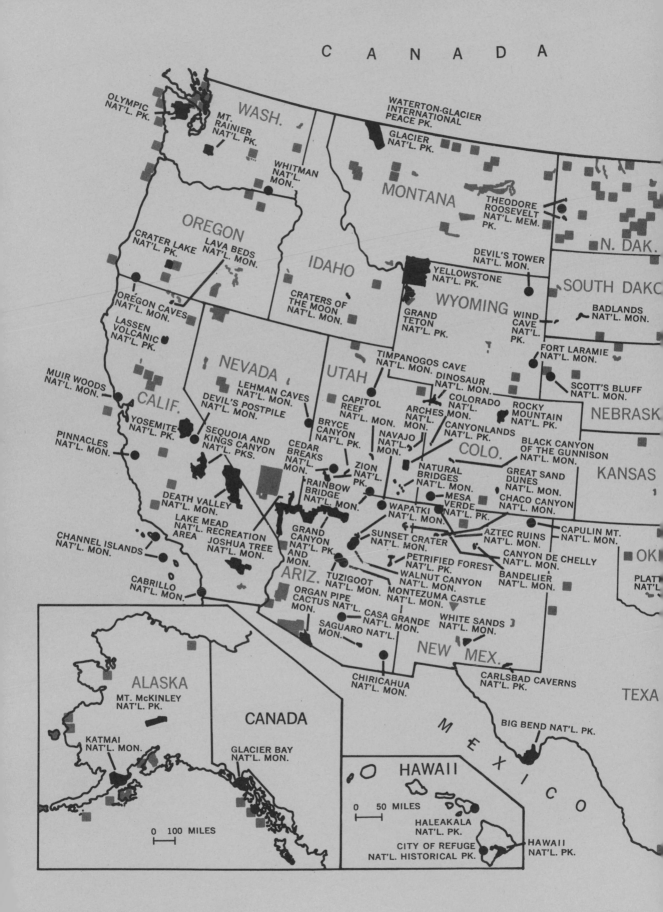

CANADA

OLYMPIC
NAT'L. PK.

WASH.

MT.
RAINIER
NAT'L. PK.

WATERTON-GLACIER
INTERNATIONAL
PEACE PK.

GLACIER
NAT'L. PK.

WHITMAN
NAT'L.
MON.

MONTANA

THEODORE
ROOSEVELT
NAT'L. MEM.
PK.

N. DAK.

OREGON

CRATER LAKE
NAT'L. PK.

LAVA BEDS
NAT'L. MON.

IDAHO

DEVIL'S TOWER
NAT'L. MON.

SOUTH DAKO

OREGON CAVES
NAT'L. MON.

CRATERS OF
THE MOON
NAT'L. MON.

YELLOWSTONE
NAT'L. PK.

WYOMING

BADLANDS
NAT'L. MON.

LASSEN
VOLCANIC
NAT'L. PK.

GRAND
TETON
NAT'L. PK.

WIND
CAVE
NAT'L.
PK.

NEVADA

UTAH

TIMPANOGOS CAVE
NAT'L. MON.

FORT LARAMIE
NAT'L. MON.

MUIR WOODS
NAT'L. MON.

LEHMAN CAVES
NAT'L. MON.

DINOSAUR
NAT'L. MON.

SCOTT'S BLUFF
NAT'L. MON.

CALIF.

DEVIL'S POSTPILE
NAT'L. MON.

CAPITOL
REEF
NAT'L. MON.

COLORADO
NAT'L.
MON.

ROCKY
MOUNTAIN
NAT'L. PK.

NEBRASK

YOSEMITE
NAT'L. PK.

SEQUOIA AND
KINGS CANYON
NAT'L. PKS.

BRYCE
CANYON
NAT'L. PK.

ARCHES
NAT'L.
MON.

CANYONLANDS
NAT'L. PK.

PINNACLES
NAT'L. MON.

CEDAR
BREAKS
NAT'L.
MON.

ZION
NAT'L.
PK.

NAVAJO
NAT'L.
MON.

COLO.

BLACK CANYON
OF THE GUNNISON
NAT'L. MON.

KANSAS

DEATH VALLEY
NAT'L. MON.

RAINBOW
BRIDGE
NAT'L.
MON.

NATURAL
BRIDGES
NAT'L. MON.

GREAT SAND
DUNES
NAT'L. MON.

LAKE MEAD
NAT'L.
RECREATION
AREA

MESA
VERDE
NAT'L. PK.

CHACO CANYON
NAT'L. MON.

CHANNEL ISLANDS
NAT'L. MON.

JOSHUA TREE
NAT'L. MON.

GRAND
CANYON
NAT'L. PK.
AND
MON.

WAPATKI
NAT'L. MON.

SUNSET CRATER
NAT'L. MON.

AZTEC RUINS
NAT'L. MON.

CAPULIN MT.
NAT'L. MON.

CANYON DE CHELLY
NAT'L. MON.

OK

CABRILLO
NAT'L. MON.

ARIZ.

TUZIGOOT
NAT'L. MON.

PETRIFIED FOREST
NAT'L. PK.

WALNUT CANYON
NAT'L. MON.

BANDELIER
NAT'L. MON.

PLAT
NAT'L

ORGAN PIPE
CACTUS NAT'L.
MON.

MONTEZUMA CASTLE
NAT'L. MON.

CASA GRANDE
NAT'L. MON.

WHITE SANDS
NAT'L. MON.

SAGUARO NAT'L.
MON.

NEW   MEX.

CHIRICAHUA
NAT'L. MON.

CARLSBAD CAVERNS
NAT'L. PK.

TEXA

ALASKA

MT. McKINLEY
NAT'L. PK.

CANADA

M E X I C O

BIG BEND NAT'L. PK.

KATMAI
NAT'L. MON.

GLACIER BAY
NAT'L. MON.

HAWAII

0   50 MILES

0   100 MILES

HALEAKALA
NAT'L. PK.

CITY OF REFUGE
NAT'L. HISTORICAL PK.

HAWAII
NAT'L. PK.

NATIONAL PARKS, MONUMENTS, AND
WILDLIFE REFUGES OF THE UNITED STATES

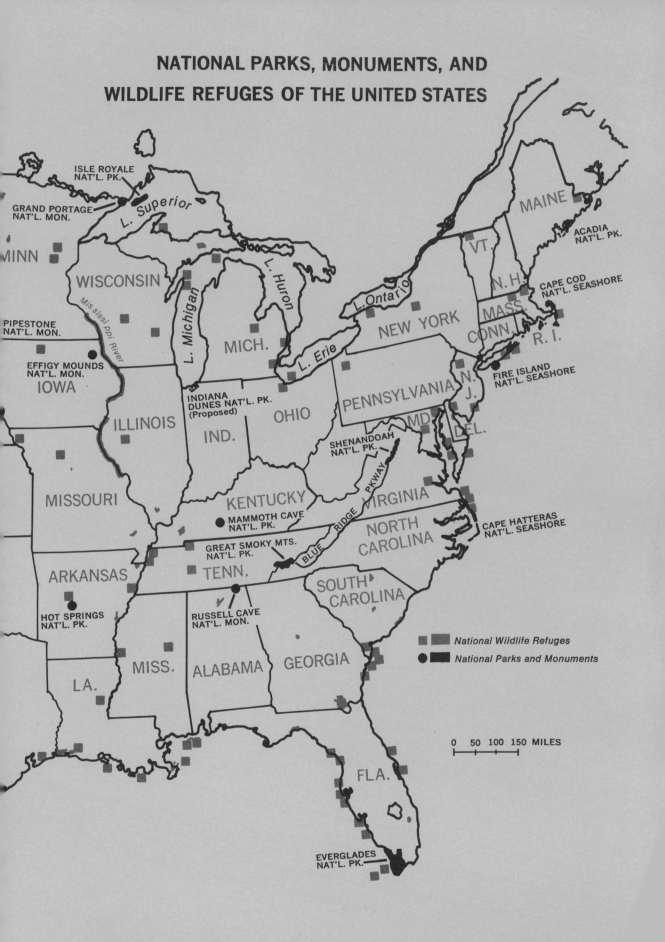

ISLE ROYALE
NAT'L. PK.

GRAND PORTAGE
NAT'L. MON.

L. Superior

MAINE

ACADIA
NAT'L. PK.

MINN

VT.

N. H.

CAPE COD
NAT'L. SEASHORE

WISCONSIN

L. Huron

L. Michigan

MASS.

CONN.

R. I.

PIPESTONE
NAT'L. MON.

MICH.

L. Ontario

NEW YORK

Mississippi River

EFFIGY MOUNDS
NAT'L. MON.

L. Erie

PENNSYLVANIA

N.
J.

FIRE ISLAND
NAT'L. SEASHORE

IOWA

INDIANA
DUNES NAT'L. PK.
(Proposed)

OHIO

MD.

DEL.

ILLINOIS

IND.

SHENANDOAH
NAT'L. PK.

MISSOURI

KENTUCKY

PKWAY

VIRGINIA

MAMMOTH CAVE
NAT'L. PK.

RIDGE

NORTH
CAROLINA

CAPE HATTERAS
NAT'L. SEASHORE

GREAT SMOKY MTS.
NAT'L. PK.

BLUE

ARKANSAS

TENN.

SOUTH
CAROLINA

HOT SPRINGS
NAT'L. PK.

RUSSELL CAVE
NAT'L. MON.

National Wildlife Refuges

National Parks and Monuments

MISS.

ALABAMA

GEORGIA

LA.

0   50 100 150 MILES

FLA.

EVERGLADES
NAT'L. PK.

with aspen (some of the groves of the latter are among the finest in the National Park System). Here also are white fir, Douglas fir, and blue spruce.

Grand Teton National Park (Wyoming)
Majestic jagged peaks of the Teton range, rising abruptly from floor of Jackson Hole. One third of the park lies above timber line. Forests are mainly of lodgepole pine, Engelmann spruce, alpine fir, and aspen. Large natural open areas on the valley floor of Jackson Hole are covered with sagebrush. Moose, wapiti ("American elk"), bighorn, and trumpeter swans may be seen here. If you are lucky enough to see a bear, admire it from a distance; bears are wild animals, not pets.

Great Sand Dunes National Monument (Colorado)
Highest inland sand dunes in U.S. In some areas, shifting dunes are burying stands of ponderosa pine. Elsewhere, as dunes move on, skeletons of dead pines buried long ago are emerging from the sand. Piñon and juniper grow here, as well as tall cottonwoods along the banks of the Medano Creek.

RHODODENDRON

Great Smoky Mountains National Park
(Tennessee, North Carolina)
Loftiest range east of the Black Hills, including many of the highest peaks in the East. The finest of eastern forests. Because northern and southern hardwood forests meet here, variety is extraordinary; 131 species of native trees are found in the park, including virgin stands of many. Largest known specimens of sixteen tree species are found in the park. Summits are cloaked with red spruce and Fraser fir, a southern extension of boreal forest zone. A group of mountain pioneer homes and farms are preserved within the park. See pages 186–191.

Isle Royale National Park (Michigan)
The largest island in Lake Superior; a rugged, roadless wilderness. Boreal conifers (white spruce, balsam fir) mingle with northern hardwoods (sugar maple, yellow birch); an outstanding example of forest transition. Successional stages are conspicuous on large area that burned in 1936. More than 200 kinds of birds, ranging from herring gull and osprey to delicate warblers. Most famous residents are the moose and timber wolves.

Lassen Volcanic National Park (California)
Lassen Peak, the only recently active volcano in the U.S. outside Alaska and Hawaii, last erupted between 1914 and 1917. Park features impressive volcanic phenomena. Peak itself is virtually barren, but three fourths of the area is forested with mixed conifers as in Yosemite, with a particularly fine stand of red

fir and Jeffrey pine. In the "Devastated Area," a gigantic mud torrent in 1915 swept aside all forests in its path; scattered pines now are becoming re-established in the area. As in all National Parks, the many wild flowers are for admiring, not for picking.

Mesa Verde National Park (Colorado)
Extensive tableland gashed by deep canyons; caves in the canyon walls shelter well-preserved prehistoric Indian cliff-dwellings. Mesa Verde (Spanish for "green table") is aptly named; the plateau is covered by dense piñon–juniper forest, an excellent example of its type. Also scrub oak, small stands of Douglas fir and ponderosa pine. Spruce Tree House was named for a group of conifers just below the ruin. The trees are actually Douglas fir; no spruce grows in the entire park.

Mount Rainier National Park (Washington)
A single peak with twenty-five glaciers radiating from its summit; most famous landmark in the Pacific Northwest. Lower slopes cloaked with dense forest of magnificent Douglas fir, western red cedar, and western hemlock. At higher elevations, silver fir, noble fir, and Alaska cedar become numerous. Subalpine meadows, studded with clumps of alpine fir and mountain hemlock, are brilliant with wild flowers throughout the summer. The difficult ascent to the snow summit is a favorite challenge among mountain climbers.

Muir Woods National Monument (California)
A virgin stand of majestic coast redwoods, the world's tallest trees; includes many over 300 feet tall. This grove occupies less than one square mile but is the only unit in the National Park System created to preserve redwoods. The trees grow only along a narrow strip of the Pacific coast in California and southern Oregon—the "fog belt." Associated trees include Douglas fir, tanbark oak, California buckeye, and others. Lush herb layer of ferns, wild flowers. See pages 151–152.

Olympic National Park (Washington)
Magnificent wilderness with rocky coastline; rugged mountains; glaciers; alpine meadows filled with wild flowers. Main attraction is the luxuriant rain forest on lower western slopes (Douglas fir; Sitka spruce; western red cedar; western hemlock; understory of bigleaf maple) which results from annual precipitation of over 140 inches. Area northeast of the park, in rain shadow of the mountains, in contrast, has the driest climate on the West Coast outside southern California (14 inches of rain per year). Resident herd of about 6000 Roosevelt elk (wapiti) is largest in the country. See pages 162–166.

WHITE-WINGED CROSSBILL

209

Pinnacles National Monument (California)

Caves, spires, and crags eroded in rocks of a volcano believed to have been active thirty million years ago. The hot, dry slopes, frequently swept by wildfire, are covered with a dense mat of evergreen shrubs (especially greasewood chamise, manzanita, and buckbrush ceanothus); considered the best example of chaparral in the National Park System. Ravens and turkey vultures seem at home in this rugged landscape.

Rocky Mountain National Park (Colorado)

Beautiful section of the Rockies, many peaks higher than 10,000 feet. Lodgepole pine most common tree below 9500 feet; but also ponderosa pine, Douglas fir, aspen, some blue spruce, and others. Engelmann spruce, alpine fir, and limber pine at higher elevations. Extensive alpine tundra. Hundreds of wild flowers. Bighorn sheep, the most spectacular resident, is often seen. Shadow Mountain National Recreation Area, adjacent to the park, is one of many such areas in the National Park System. Managed primarily as an outdoor playground.

Sequoia and Kings Canyon National Parks (California)

SUGAR PINE

Two adjacent parks in the Sierra Nevada; Mount Whitney in Sequoia National Park is the highest peak in U.S. except for Alaska (14,495 feet). Many groves of giant sequoias, which here reach their greatest size. Other prominent trees of middle elevations are ponderosa pine, sugar pine, white fir, incense cedar, and California black oak. (Sugar pine is at its best here and at Yosemite.) Major trees of higher slopes include red fir, Jeffrey pine, lodgepole pine, western white pine, with whitebark pine and foxtail pine near timber line. Beautiful alpine meadows below the snow-capped peaks. On the lower western slopes are fine examples of typical chaparral vegetation made up of many species of evergreen shrubs and of oak woodland.

Shenandoah National Park (Virginia)

Scenic area of Blue Ridge Mountains traversed by famous Skyline Drive. Magnificent vistas; wealth of wild flowers. Extensive Appalachian oak forests; hemlock on cooler slopes; red spruce and balsam fir on crests. Gaunt remains of chestnut trees, killed by chestnut blight in 1930s, are common, provide dens and nest sites for many animals. Since the area was cut for lumber and farming, few of the forests are virgin; young forests creeping across abandoned fields are a common sight.

Wind Cave National Park (South Dakota)

Features an unusual limestone cavern in the Black Hills with boxwork formations. Half the aboveground area is rolling prairie, with bison, pronghorn, and prairie dogs; other half is forested,

210

mainly with ponderosa pine. A biological meeting ground—ponderosa pine and Rocky Mountain juniper are typical of western mountains; American elm and bur oak are eastern; yucca and cottonwood are characteristic of arid southwest.

Yellowstone National Park (Wyoming, Montana, Idaho)
Prime attractions are the 3000 geysers and hot springs (including Old Faithful), and the spectacular falls and canyon of the Yellowstone River. Ninety per cent of the park is forested, primarily with lodgepole pine; alpine fir and Engelmann spruce at higher elevations. Quaking aspen is the pioneer on burned areas. Like all the National Parks, Yellowstone is a sanctuary for wildlife, including the white pelican, the rare and beautiful trumpeter swan, black and grizzly bear, pronghorn, bison, bighorn sheep, wapiti, moose, and many others. See pages 170–173.

WHITE PELICAN

Yosemite National Park (California)
Beautiful mountain region of the Sierra Nevada. Prime attractions are three groves of giant sequoias; Yosemite Valley, a glacier-cut gorge with sheer granite walls and spectacular waterfalls. Most of the park is high mountain wilderness, with mixed conifer forests typical of the Sierra Nevada region. Conspicuous mountain zonation, ranging from chaparral in the dry foothills to alpine tundra at the summits. Generally the same trees as those in Sequoia and Kings Canyon National Parks, except for foxtail pine.

ANIMAL TRACKS

The tracks of almost every kind of animal are distinctive. Identifying characteristics include track patterns, size and shape, number of toes on each foot, and presence or absence of clawmarks or tailprints. By following tracks in snow, mud, dust, or sand and by noticing other bits of evidence such as feeding signs, it is possible to observe the habits and ecology of elusive, seldom-seen animals. The tracks of several common animals are pictured here. The larger tracks show details of the imprint made by the right front foot (left) and the right hind foot (right) of each. The smaller prints indicate the typical pattern of tracks left by the animal when walking or hopping at slow speed.

BOBCAT

4"

DEER MOUSE

⁷⁄₈"

STRIPED SKUNK

2½"

COTTONTAIL

4"

RED SQUIRREL

1¾"

RED FOX

1¾"

RACCOON

4¼"

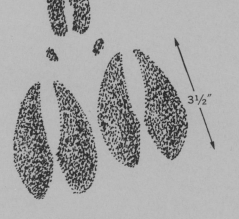

WHITE-TAILED DEER

3½"

OPOSSUM

2"

BLACK BEAR

9"

Endangered Forest Animals

In the few centuries since the colonization of America, an appalling number of animal species have slipped—or have been pushed—into the oblivion of extinction. While man has not been responsible for the extermination of every species, in an alarming number of cases his efforts to remold the face of the continent have had tragic effects on wildlife. Many species simply have been unable to survive the destruction of their age-old habitat. Others have been shot, trapped, poisoned, or driven ever deeper into the wilderness, either because man would not tolerate them near his homes and farms or for the sake of their fur, flesh, or feathers. More recently, man has begun polluting his environment with pesticides—DDT, 2-4-D, and a host of others—whose sinister effects on wildlife we are only beginning to understand.

Although it is too late to undo all the tragic errors of the past,

FLORIDA KEY DEER

This subspecies of the familiar white-tailed deer lives only on the island keys off southern Florida. About the size of a large dog, it is America's smallest deer. As a result of hurricanes, fires, and excessive hunting, fewer than fifty remained alive in 1947, but vigorous protection and the establishment of refuge lands have brought the herd back to approximately 200. Even so, the future of this appealing miniature deer remains uncertain. If you are lucky, you may glimpse the key deer while driving across Big Pine Key.

TIMBER WOLF

At one time found through most of North America, the timber wolf has been relentlessly persecuted by settlers and ranchmen as civilization advanced across the continent. Only in western Canada and Alaska, in Isle Royale National Park, Michigan, and in remote forests of northern Michigan and Wisconsin can you now expect to hear the stirring nighttime howl of this magnificent symbol of wilderness.

many things can be done to ensure the survival of our remaining wildlife. By controlling the hunting, fishing, and trapping of more abundant species, we can assure the survival of adequate breeding stocks. We can preserve areas for wild animals to live in and can provide wildlife habitat on lands used primarily for other purposes. But most of all we must change our attitudes toward wild creatures and respect their right to share our environment.

Four forest species whose numbers have ebbed to dangerously low levels are listed below. Whether or not they will continue to delight future generations as living remnants of America's wilderness heritage depends on the protection and encouragement they receive now and in the immediate future.

WOLVERINE

The rare and secretive wolverine is a fast-moving, powerfully built hunter about the size of a bear cub. A symbol of fearless stealth and cunning, it can kill animals as large as deer and caribou, yet is also satisfied to steal the bait and captured animals from woodsmen's trap lines. Formerly found throughout the northern border states and south along western mountain ranges, the few remaining in the United States are restricted to wilderness tracts in Alaska, Glacier National Park, Montana, and parts of the Sierra Nevada range. Count yourself fortunate if you discover even its tracks.

IVORY-BILLED WOODPECKER

Largest and most striking of North American woodpeckers, the ivory-bill is so rare it may already be extinct. Though never abundant, it once occurred throughout the vast riverbottom forests of the Southeast; logging, fires, and the relentless press of civilization have destroyed most of its habitat. Because its diet is restricted almost exclusively to certain insect larvae that live between the bark and sapwood of trees that have been dead no longer than two years, the ivory-bill requires vast tracts of wilderness to survive; dead trees of the proper age are widely scattered in the forest. Only by preserving large areas of southeastern timberland can we offer the rare ivory-billed woodpecker even the slimmest chance for survival.

Stories in Stumps

Fire scar — new growth is gradually covering the wound

One conspicuous feature on any freshly cut tree-stump surface is a series of concentric circles of dark and light wood. By learning to interpret clues recorded in these rings, you can uncover many details of the tree's history and of the forest in which it grew.

Annual rings, as these circles are called, result from the fact that trees grow in girth as well as in height. Every year a new layer of wood forms just beneath a tree's bark. Like a sheath extending up the trunk and around every branch, the layer of new wood envelops the layer of wood formed the preceding year. The lighter portion of each ring represents rapid growth early in the season; the narrower, darker portion indicates the end of the growing season. Although a few kinds of trees may form more than one distinct layer of wood during a single year, and in an extremely unfavorable growing season no wood at all, these are exceptions. As a general rule, each annual ring equals one year of a tree's life.

You can therefore determine the age of a tree when it was cut simply by counting the annual rings. If you know the exact year it was cut, you can assign a specific date to events recorded in any annual ring in the tree's lifetime. You can even date fairly precisely the year in which it began to grow. These calculations will be exact only if the stump was cut at or near ground level. If the tree was cut well above the ground, the record of its earliest years will not be visible. Such a gap in the tree-ring record may be serious; trees growing under unfavorable conditions such as heavy shade or seedlings repeatedly browsed by

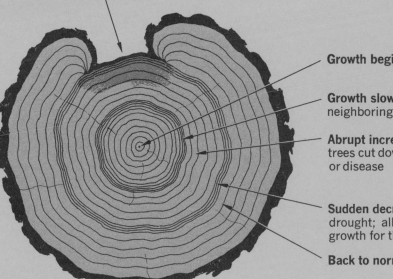

Growth begins

Growth slows — increasing competition from neighboring trees

Abrupt increase in growth rate — neighboring trees cut down or damaged, as by wind storm or disease

Sudden decrease in growth rate — probably drought; all stumps in the area show little growth for these years

Back to normal

deer or cattle may require several seasons to grow as much as a foot tall.

Even more revealing than the number of rings are variations in individual ring width that indicate changes in a tree's rate of growth. Such factors as the availability of light and water or insect damage can cause drastic changes in the amount of wood laid down in a single year. One tree, growing under extremely favorable conditions, may increase an inch or more in diameter in a single year; another, growing on rocky mountain slopes near timber line, may require ten years to increase even one tenth of an inch in diameter. By analyzing variations in the width of successive annual rings, then, it is possible to piece together events in the tree's history. If growth has been very slow, you may need a magnifying glass to distinguish all the annual rings.

Recently cut stumps are easiest to read. When a stump has been exposed to the weather for several seasons, the tree-ring record becomes blurred. In such a case, smoothing the surface with sandpaper will usually make the rings show up more distinctly. Or, if the wood is not too hard, a V-shaped groove can be cut into the surface of the stump, with the cut in a straight line from the center to the edge of the stump. The rings will stand out clearly along the unweathered surface of the fresh cut.

To make a permanent record of a stump's tree-ring history, place and tack a strip of thin white paper from the center of the stump to the bark at the edge. Rub lightly on the paper surface with the broad edge of the point of a soft-lead pencil. The annual rings will show up on the paper as dark lines alternating with lighter areas. Marking the dark lines with ink will provide a convenient, durable record of the tree's growth history.

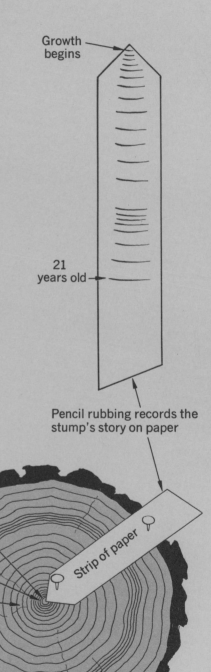

Growth begins

21 years old

Pencil rubbing records the stump's story on paper

Strip of paper

Growth begins

Slow growth — competition of neighbors

Gradual increase as tree overtops neighbors

Sudden decrease — probably insect epidemic; only stumps of this species show decrease for these years; other species grew normally

Back to normal

How to Make a Terrarium

A terrarium is a miniature garden in a sealed glass container. It is an ideal place to preserve a living sample of the forest floor, either for purely decorative purposes or as a convenient laboratory where you can observe the habits of things that live on the forest floor. Any clear glass container is suitable. A leaky aquarium is excellent. A wide-mouth gallon jar makes an inexpensive terrarium; use it upright or cradle it on its side on a wooden base. Or make one from window panes. Standard 8 × 10-inch single-strength window glass makes a convenient-sized terrarium. Cut two inches off each end piece, or ignore them and let them project above the top of the terrarium (see diagram).

You can begin by lining the bottom of the container with a layer of marble-sized pebbles in order to provide drainage; then add a layer of soil. Don't waste space with a thick bed of soil; you need only enough to anchor the plants in place. (For a more interesting effect, bank the soil toward the rear of the container.) Ordinary garden soil is unsatisfactory because it tends to become waterlogged. Woodland plants grow better in a mixture of humus and soil from the forest in which they are collected. As a bonus, the forest soil will be populated with an interesting assortment of sowbugs, mites, insects, and other animals.

Almost any small plant growing on the forest floor, including tree seedlings, is suitable for use in a terrarium. For best results, select plants from similar habitats with similar growth requirements. Don't try to crowd too many into the limited space; use only small plants and leave enough room for them to grow. Press the roots gently into place and cover them lightly with soil. You

PIPSISSEWA

SPRING PEEPER

LAND SNAIL

COMMON POLYPODY

WINTERGREEN

will create a more attractive effect by placing larger plants to the rear and smaller ones in the foreground. Fill in blank spaces with pads of moss and accent the arrangement with lichen-crusted stones, a bright red fungus, snags of wood, or pieces of bark—things you might expect to find on the woodland floor.

When all the plants are in place, moisten the terrarium with a syringe or laundry bulb-sprinkler, so that it is moist but not soggy, and clean the sides with a piece of tissue. Don't over-water. If puddles form at the bottom, allow the water to evaporate for a few days. Finally, seal the container with a close-fitting glass cover or with a sheet of transparent plastic held in place by an elastic band. Prepared this way, the terrarium will be self-sustaining for many months before requiring water. Water lost by the plants through transpiration will condense on the sides and cover of the container and trickle back to the soil to be reused again and again by the plants and animals. Do not keep the terrarium in full sunlight—too much heat and light will kill the plants. Their natural habitat is the forest floor, where they thrive in heavy shade.

If you wish, add a small toad, a salamander, a couple of land snails or other inhabitants of the forest floor. Provide hiding places beneath flat stones or pieces of bark, and keep the animals supplied with earthworms, insects, or whatever is their natural food. Many animals such as salamanders and toads rarely survive the winter in captivity because it is so difficult to provide their natural food then. It is best to release them in their natural habitat in autumn, early enough for them to hibernate.

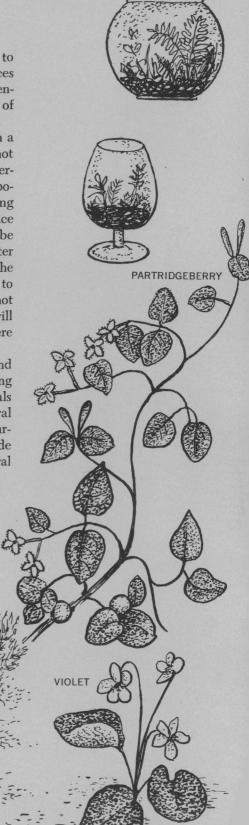

PARTRIDGEBERRY

GOLDTHREAD

TIGER SALAMANDER

VIOLET

RED EFT

How to Make a Plant Collection

As you learn to identify the trees and other plants of your area, you may decide to make a collection of their leaves, twigs, blossoms, and fruits. A beginning collection could consist simply of leaves pressed between the pages of a heavy book and then mounted on cards. The results will be far more durable and attractive, however, if specimens are gathered and prepared with greater care. Instead of collecting at random, concentrate on trees and shrubs of a particular area. Ferns and other plants of the forest floor could also be included.

A specimen from a tree is not complete unless it includes enough of the twig to demonstrate the arrangement of leaves on the stem. Select a twig several inches long with at least two or three leaves on it and cut it from the branch with a sharp knife. Gather imperfect as well as perfect leaves; galls, the tunnelings of leaf miners, and other signs of insect life represent important facets of the tree's life history. Collect from the same tree at different seasons so that your collection will include specimens of both blossoms and fruits. If you store the specimens in plastic bags until you get home, they will remain fresh for several hours.

Number each specimen as it is gathered, either by storing each in a separate numbered bag or by tying an identifying tag to each twig. Keyed to a corresponding number in your notebook, record anything about the plant that is not obvious from the specimen. Habitat, color, and texture of bark on the trunk should be noted, as should form of the tree, height, abundance in the area, location, and the identity of associated trees.

Although specimens can be kept a day or two if they are stored in a cool place, it is better to press them as soon as possible. A plant press can be as simple or as complex as you care to make it. The basic necessities are absorbent paper and a pressure source. Newspaper and desk-size blotters are ideal for absorbing moisture as the plants dry. Crease a sheet of newsprint in the form of a folder and place it on a blotter or a stack of three or four more sheets of paper. Inside the folder, arrange the plant specimen in the desired position. Spread the leaves so that they do not overlap, and turn some of them over to show their undersides. If necessary, discard a few of the leaves. Close the folder and place several more layers of paper or a blotter on top, then

PLANT PRESS

spread the next specimen in another folder. Continue to add papers and specimens layer by layer. If many plants are being pressed at one time, insert a sheet of corrugated cardboard after every three or four specimen folders so that air can circulate between the layers. Each specimen folder should be numbered for identification. When all the plants have been arranged, sandwich the entire stack between two sheets of plywood or some other wood and bind it tightly with two straps or pieces of heavy cord. Pressure will prevent the leaves from rumpling as they dry.

Store the entire plant press a few feet from a stove or furnace or in some other warm, dry place. After about twenty-four hours, replace the newspapers and blotters with dry ones, but leave the specimens in the same sheets, so that you do not disturb their positions. Strap the press shut again and let the plants continue to dry for a week or ten days, changing the papers every two or three days. The used sheets will dry out and can be used again.

To preserve apples and other fleshy fruits, cut a thin slice across the center so that the seeds and their arrangement are visible. Dry the slice in the plant press with the leaf specimen. Berries and other small fruits will dry satisfactorily without special treatment. Nuts and pine cones need not be dried; simply store them in small boxes or jars. As soon as spruce and hemlock boughs have dried enough to lie flat, dip them in shellac, otherwise the needles will fall off.

When the specimens are thoroughly dry, remove them from the press and mount them on sheets of heavy white paper or thin cardboard (the standard size used by botanists is 11½ by 16½ inches). Glue the specimen in place or fasten it with strips of gummed tape. Each sheet should include a label stating the name of the plant, name of the collector, and the date and place it was collected. If you are keeping a catalogue of your collection and a notebook with additional information about the plant, number the mounted specimen. Place each completed sheet in a manila folder for protection, and store the folders in boxes in a dry place. A few moth flakes in each box will prevent insect damage. Properly prepared and handled with care, a plant specimen will remain in good condition for many years.

SPECIMEN SHEET

FORESTS FOR MAN

From their foliage to their roots, trees yield an astonishing variety of products. More than 5000 different things used every day originate in the 509 million acres of commercially productive forest land in the United States. A few of the multitude of forest products are listed below.

FOREST PRODUCTS

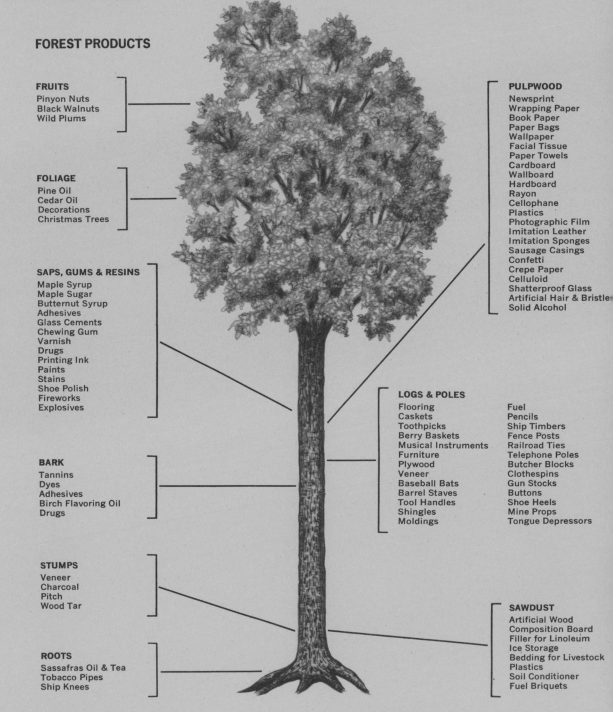

FRUITS
Pinyon Nuts
Black Walnuts
Wild Plums

FOLIAGE
Pine Oil
Cedar Oil
Decorations
Christmas Trees

SAPS, GUMS & RESINS
Maple Syrup
Maple Sugar
Butternut Syrup
Adhesives
Glass Cements
Chewing Gum
Varnish
Drugs
Printing Ink
Paints
Stains
Shoe Polish
Fireworks
Explosives

BARK
Tannins
Dyes
Adhesives
Birch Flavoring Oil
Drugs

STUMPS
Veneer
Charcoal
Pitch
Wood Tar

ROOTS
Sassafras Oil & Tea
Tobacco Pipes
Ship Knees

PULPWOOD
Newsprint
Wrapping Paper
Book Paper
Paper Bags
Wallpaper
Facial Tissue
Paper Towels
Cardboard
Wallboard
Hardboard
Rayon
Cellophane
Plastics
Photographic Film
Imitation Leather
Imitation Sponges
Sausage Casings
Confetti
Crepe Paper
Celluloid
Shatterproof Glass
Artificial Hair & Bristles
Solid Alcohol

LOGS & POLES
Flooring   Fuel
Caskets   Pencils
Toothpicks   Ship Timbers
Berry Baskets   Fence Posts
Musical Instruments Railroad Ties
Furniture   Telephone Poles
Plywood   Butcher Blocks
Veneer   Clothespins
Baseball Bats   Gun Stocks
Barrel Staves   Buttons
Tool Handles   Shoe Heels
Shingles   Mine Props
Moldings   Tongue Depressors

SAWDUST
Artificial Wood
Composition Board
Filler for Linoleum
Ice Storage
Bedding for Livestock
Plastics
Soil Conditioner
Fuel Briquets

Alpine zone: The portion of a mountain that lies above timber line. *See* Tundra; Vegetation zones, vertical.

Annual: A plant that completes its life cycle from seedling to mature seed-bearing plant during a single growing season, then dies.

Anthocyanins: A group of pigments that produces shades of red and blue in leaves.

Association: A grouping of plants and animals that repeatedly occur together in a forest region. Associations may be identified in terms of their predominant tree associates, as oak–hickory forest.

Biennial: A plant that lives for two growing seasons, producing only leaves during the first season, flowers and seeds during the second.

Biological clock: A time-measuring process within a living organism. The exact nature and location are still unknown, but biological clocks govern responses to many external events, such as the changes in the length of day from season to season.

Bog: A wet, low area, often an old lake bed, filling or filled with partially decayed matter known as peat.

Boreal forest: Northern conifer forest.

Botany: The scientific study of plants.

Broadleaf: The term describing a plant with wide-bladed leaves, such as an oak or maple; generally refers to flowering trees in contrast to conifers.

Browse: To eat the twigs and leaves of woody plants. Deer, moose, and their relatives are browsers.

Canopy: Layer formed by the leaves and branches of the forest's tallest trees.

Carotenoids: A group of pigments that produces yellow, orange, and red hues in plants.

Chaparral: Dense scrub vegetation of broadleaf evergreen or wintergreen shrubs.

Chinook: A warm, dry wind that blows down the leeward slopes of mountains.

Chlorophyll: A group of pigments that produces the green hue of plants; essential to *photosynthesis*.

Chrysalis: The hard-shelled *pupa* of a butterfly.

Climate: The average weather conditions of an area, including temperature, rainfall, humidity, windiness, and hours of sunlight, based on records kept for many years.

Climax: The relatively stable association that represents the final stage in a *sere* under existing conditions of soil, climate, and human interference.

Cocoon: The silky case that protects certain insects, such as moths, during their pupal stage.

Community: All the plants and animals in a particular habitat that are bound together by *food chains* and other interrelations.

Cone: A structure composed of many spirally arranged scales in which pollen or ovules are produced. Cones differ from flowers in that the ovules are borne on the surfaces of the scales, or carpels. In a flower the carpels form a container called the *pistil*, inside which the ovules are borne.

Conifer, Coniferous: A plant that bears its seeds in cones. Usually refers to needleleaf trees, although some, such as yew, do not bear cones.

Conservation: The use of natural resources in a way that assures their continuing availability to future generations; the intelligent use of natural resources.

Corm: A thick, bulblike, underground plant stem.

Deciduous: Term describing a plant that periodically loses all its leaves, usually in autumn. Most North American broadleaf trees are deciduous. A few conifers, such as larch and cypress, also are deciduous. *See* Evergreen.

Decomposer: A plant or animal that feeds on dead material and causes its mechanical or chemical breakdown.

Dendrology: A branch of *botany* devoted to the study of trees.

Desert: An unforested area where the rainfall is very slight and unevenly distributed throughout the year and where the daytime temperatures are high. The sparse vegetation is composed of shrubs, many kinds of annual plants, and grasses; in some desert areas cacti are common.

Ecology: The scientific study of the relations of living things to one another and to their environment. A scientist who studies these relationships is called an *ecologist*.

Ecosystem: All living things and their environment in an area of any size. All are linked together by energy and nutrient flow.

Environment: All the external conditions surrounding a living thing.

Evergreen: A plant that does not lose all of its leaves at one time. Among trees, some broadleaf species, such as live oak, remain green all year, but most North American evergreens are coniferous. *See* Conifer; Deciduous.

Flower: A plant structure that is usually composed of petals, stamens, and one or more pistils. Ovules and seeds are produced within the pistil. *See* Cone.

Food chain: A series of plants and animals linked by their food relationships. A green plant, a leaf-eating insect, and an insect-eating bird would form a simple food chain. Any one species is usually represented in several or many food chains.

Foothill zone: The lowest band of vegetation in a mountainous region that differs from the vegetation of surrounding lowlands.

Forest: A complex community of plants and animals in which trees are the most conspicuous members.

Forest floor: The layer of decomposing material that covers the soil in a forest.

Forest region: An extensive area of a continent in which the *climax*-forest associations are closely similar. The major forest regions of North America are western, northern or boreal, eastern, and subtropical.

Form: A small protected place in which an animal, especially a rabbit or hare, rests.

Gall: An abnormal growth resulting from chemical or mechanical irritation of plant tissue. Most galls are caused by insects.

Girdling: Stripping or gnawing a section of bark around the trunk of a tree or shrub; may eventually kill the plant.

Grassland: A vegetation type in which grasses are the most conspicuous plants.

Habitat: The immediate surroundings (living place) of a plant or animal.

Hardwood: A deciduous or broadleaf tree. *See* Softwood.

Herb: Any flowering plant or fern that has a soft, rather than woody, stem.

Herb layer: The layer of soft-stemmed plants growing close to the forest floor.

Hibernation: A prolonged dormancy or sleeplike state in which animal body processes slow down drastically and no food is eaten. Nearly all cold-blooded animals and a few warm-blooded animals hibernate during the winter.

Larva (plural *larvae*): An active, immature stage in the life history of an insect, such as the caterpillar stage in the development of a butterfly.

Leader: The main shoot growing from the top of a tree with a single main trunk.

Mast year: A year of above-average nut production in a forest.

Microclimate: "Little climate"; the environmental conditions in a restricted area.

Microhabitat: A "small habitat" within a larger one in which environmental conditions differ from those in the surrounding area. A hole in a tree trunk or an animal carcass is a microhabitat within the forest.

Mixed forest: A forest that includes both coniferous and deciduous trees.

Montane zone: The band of vegetation that occurs at intermediate elevations in mountainous regions between foothill and subalpine zones. *See* Vegetation zones, vertical.

Mor: A type of forest floor formed by a thick mat of slowly decomposing matter, often conifer needles.

Mull: A type of forest floor and soil in which the decomposing matter, usually formed of broad leaves, decays rapidly. The humus is mixed thoroughly with mineral matter by earthworms and other small animals, so there is no sharp boundary between the forest floor and soil.

Muskeg: A mossy *bog* in the northern coniferous forest region.

Needleleaf: Bearing needlelike leaves. *See* Conifer.

Old field: Farm land once cultivated, but now untended.

Ovulate flower parts: The "female," or seed-producing parts of a flower; the *pistil* and its contents. *See* Ovule.

Ovule: The structure which contains the female germ cell (egg) and which develops into a seed. *See* Cone; Flower.

Parasite: A plant or animal that lives in or on another living thing (its host) and obtains part or all of its food from the host's body.

Perennial: A plant that lives for several years and usually produces seeds each year.

Photosynthesis: The process by which green plants convert carbon dioxide and water into simple sugar. *Chlorophyll* and sunlight are essential to the series of complex chemical reactions involved.

Pigment: A chemical substance that reflects or transmits only certain light rays and thus imparts color to an object. For example, a substance that absorbs all but red rays, which it reflects, will appear red. *See* Anthocyanin; Chlorophyll; Tannins.

Pistil: The "female" structure of a flower in which the ovules are produced. The ovules, when fertilized by male cells from pollen, develop into seeds. *See* Flower; Ovule; Pollen.

Pollen: Tough, minute, grainlike bodies produced in the stamens of flowers or in staminate conelets which contain the male germ cell. Blown or carried to a pistil, the pollen grain develops a tubelike outgrowth which penetrates to an ovule. The germ cell moves through the tube into the ovule, and fertilization occurs. The ovule then develops into a seed. *See* Cone; Flower; Ovule.

Predator: An animal that lives by capturing other animals for food.

Pupa (plural *pupae*): The inactive stage in an insect's life history when the *larva* is transforming into an adult. *See* Chrysalis; Cocoon.

Rain shadow: An area on the leeward side of a mountain barrier that receives little rainfall.

Rhizome: A horizontal rootlike stem that grows in the soil.

Savanna: A parklike grassland with scattered trees or clumps of trees.

Scavenger: An animal that eats the dead remains and wastes of other animals and plants.

Scrub: A low woody vegetation composed principally of shrubs.

Seral stage: One community of a *sere*.

Sere: The series of communities that follow one another in a natural succession, as in the change from a bare field to a mature forest.

Shrub: A woody plant less than twelve feet tall, usually with more than one stem rising from the ground.

Softwood: A coniferous tree. A common but not strictly accurate term; the wood of many conifers is harder than that of some so-called hardwood trees.

Staminate flower parts: The "male," or pollen-producing, parts of a flower; the stamens. *See* Flower; Pollen.

Stomate: A microscopic opening in the surface of a leaf that allows gases to pass in and out.

Subalpine zone: The band of vegetation in mountainous regions that occurs below timber line and alpine zone. *See* Vegetation zones, vertical.

Succession: The gradual replacement of one community by another. *See* Sere.

Tannins: A group of pigments that produce brown leaf coloration.

Territory: An area defended by an animal against others of the same species. Used for breeding, feeding, or both.

Timber line: The upper limit of tree growth on mountains. A band of stunted and usually oddly shaped trees between the subalpine forests and alpine tundra. *See* Vegetation zones, vertical.

Transpiration: The process by which water evaporates from plant tissues.

Tree: A woody plant twelve or more feet tall with a single main stem (trunk) and a more or less distinct crown of leaves.

Tundra: Treeless vegetation in regions with long winters and low annual temperatures. Arctic tundra extends north of the *boreal forests*. Alpine tundra extends above *timber line* on mountains.

Understory: The layer formed by the crowns of smaller trees in a forest.

Vegetation: The mass of plants that covers a given area. Flora, a term often wrongly used interchangeably, is a list of the species of plants that compose the vegetation.

Vegetation zones, vertical: The horizontal belts of distinctive plant cover in mountainous regions, resulting from climatic changes related to elevation changes. From base to peak, the zones are foothill, montane, subalpine, timber line, and alpine.

Winter-bare forest: A forest composed of deciduous trees.

Yard up: To gather in a sheltered area in winter; used in reference to deer, moose, and their relatives.

Illustration Credits and Acknowledgments

COVER: Mule deer, Kaibab Forest, Arizona, Sonja Bullaty.

UNCAPTIONED PHOTOGRAPHS: 8–9. Redwood trees, California, Victor B. Scheffer. 70–71. First snow in Yosemite National Park, California, Robert Barbee. 136–137. Cypress forest from the air, South Carolina, Walter Dawn.

ALL OTHER ILLUSTRATIONS (*credits are separated from top to bottom by a colon and from left to right by a semicolon*): 10. Henry C. Johnson: Robert B. Smith 11. Dr. Alexander B. Klots: Wilford L. Miller 12. John H. Gerard 13. Jack Boucher, from National Park Service: United States Department of the Interior 14. Charles Fracé 15. Larry Pringle: Roche 16. Larry Pringle 17. Felix Cooper 18. Robert W. Mitchell: Dr. Alexander B. Klots 19. Bill Ratcliffe 20. Graphic Arts International 21. Tommy Lark from Photo Trends 22. Dr. Alexander B. Klots 23. Richard J. Scheich 24. Walter Dawn 25. Graphic Arts International 26. Leonard Lee Rue III: Dr. Alexander B. Klots 27. Grambs Miller 28. B. B. Jones: Michael Wotton 29. Torrey Jackson 30. Ansel Adams 31. Jack Dermid 32. Thase Daniel 33. Victor B. Scheffer: Roche 34. Dr. William M. Harlow: Walter Dawn 35. Dr. William M. Harlow 36–37. Felix Cooper 38. Dr. Alexander B. Klots 39. Shostal Agency 40. Ansel Adams 41. Michael Wotton 42. U. S. Forest Service 43. Leonard Lee Rue III 44–46. Patricia Hendricks 47. Sonja Bullaty 48. Graphic Arts International 49. Dr. Richard B. Fischer 50–55. Patricia Hendricks 56–57. Bill Ratcliffe 58. Stephen Collins 59. Betty Binns 60–61. Graphic Arts International 62. Stephen Collins 63. Les Line, *Audubon Magazine* 64–65. Sonja Bullaty 66–67. U. S. Forest Service 68. Sonja Bullaty 72–73. Felix Cooper 74. Leonard Lee Rue III: Ernest Gay 75. J. M. Conrader: Dr. Richard B. Fischer 76. Jeanne White 77. William H. Sager; William W. Dunmire: Robert W. Carpenter 78. Roche 79. Graphic Arts International 80. Henry C. Johnson 81. Glenn D. Chambers 82. Glenn D. Chambers 83. Jack Dermid: J. M. Conrader 84. Glenn D. Chambers 85. Torrey Jackson 86. Matthew Vinciguerra 87. Jack Dermid 88–90. Dr. William M. Harlow 91. Patricia Hendricks 92. Luoma Photos 93. Hugh Spencer 94–95. Felix Cooper 96. Grant Haist 97. J. M. Conrader 98. Dr. C. J. Stine 99. John H. Gerard from National Audubon Society 100. G. Ronald Austing from National Audubon Society 101. Hugh M. Halliday from National Audubon Society 102. Dr. Alexander B. Klots: Leonard Lee Rue III: John H. Gerard 103. Wilf Taylor 104. Bill Reasons 105. Graphic Arts International 106. Walter Dawn 107. H. D. Wheeler 108. Dr. Alexander B. Klots 109. Jack Dermid 110–111. Allan D. Cruickshank from National Audubon Society 112. Roche 113. Larry Pringle 114–115. Sonja Bullaty 116. Patricia Hendricks 117. Sonja Bullaty 118. Peter G. Sanchez 119. Robert C. Hermes from National Audubon Society 120. Wilford L. Miller 121. Ed Cesar 122. Harry L. Beede 123. Patricia Hendricks 124. Bob and Ira Spring 125. Graphic Arts International 126. Leonard Lee Rue III 127. Verna R. Johnston 128. Nick Drahos from New York State Conservation Department 129. Leonard Lee Rue III from National Audubon Society 130–131. Charles Fracé 132. Dr. Richard B. Fischer 133. Leonard Lee Rue III 134. Bill Ratcliffe 138. Graphic Arts International 139. Willis Peterson: National Park Service Photo 140. Felix Cooper 141. William Garnett 142–143. Graphic Arts International 144–145. Charles Fracé 146. Felix Cooper 147–148. Howard King, Save-the-Redwoods League 149. Jim Yoakum from Monkmeyer Press Photo 150. Wayne W. Bryant from National Park Service 151. Robert Barbee 152. Charles Fracé 153. Robert Barbee 154–155. Ansel Adams 156. V. E. Ward 157. Verna R. Johnston 158. Charles Fracé 159. Ansel Adams 160. Wilford L. Miller 161. Bill Ratcliffe 162. Bob and Ira Spring 163. Charles Fracé 164–165. Ansel Adams 166. Ruth Kirk 167. Graphic Arts International 168. Wayne W. Bryant 169. Felix Cooper 170. Charles Fracé 171. Allan D. Cruickshank from National Audubon Society 172. Willis Peterson 173. Ed Park 174–175. Dr. John Marr 176. Freeman Heim from U. S. Forest Service 177. Annan Photos 178. John H. Gerard 179. Charles J. Ott 180–181. Willis Peterson 182. Ernest Gay 183. Grambs Miller 184. Ed Cesar 185. Jack Dermid 186–187. Ansel Adams 188. Luoma Photos 189. Thase Daniel 190. David Morhardt from National Audubon Society 191. Irvin L. Oakes 192. Grant Haist 193. Max Hunn 194. Max Hunn from Annan Photos 195. Max Hunn 196–197. George Raz 198. Howard King, Save-the-Redwoods League 204–205. Charles Fracé 208–215. Charles Fracé 216–217. Felix Cooper 218–221. Charles Fracé 222. Felix Cooper

PHOTO EDITOR: ROBERT J. WOODWARD

ACKNOWLEDGMENTS: *The publisher is grateful to Doubleday & Company, Inc., for permission to reprint a short excerpt from* My Wilderness: The Pacific West *by William O. Douglas, copyright © 1960 by William O. Douglas. The publisher also wishes to thank Wayne W. Bryant and William Perry of the National Park Service, who read the entire manuscript and offered valuable suggestions; Audubon Magazine, L. J. Prater and E. Larson of the U.S. Forest Service, and M. Woodbridge Williams of the National Park Service for their assistance in locating photographs.*

Bibliography

FORESTS AND TREES

BRAUN, E. LUCY, *Deciduous Forests of Eastern North America*. Blakiston, 1950.

COLLINGWOOD, G. H., and WARREN D. BRUSH, *Knowing Your Trees*. American Forestry Association, 1964.

FARB, PETER, and THE EDITORS OF LIFE, *The Forest*. Time, Inc., 1961.

HARLOW, WILLIAM M., *Trees of the Eastern and Central United States and Canada*. Dover, 1957.

KITTREDGE, JOSEPH, *Forest Influences*. McGraw-Hill, 1948.

MC CORMICK, JACK, *The Living Forest*. Harper & Row, 1959.

MC MINN, HOWARD E., and EVELYN MAINO, *An Illustrated Manual of Pacific Coast Trees*. University of California Press, 1956.

PEATTIE, DONALD CULROSS, *A Natural History of Trees of Eastern and Central North America*. Houghton Mifflin, 1950.

PEATTIE, DONALD CULROSS, *A Natural History of Western Trees*. Houghton Mifflin, 1953.

PETRIDES, GEORGE A., *A Field Guide to Trees and Shrubs*. Houghton Mifflin, 1958.

U.S. DEPARTMENT OF AGRICULTURE, *Trees, The Yearbook of Agriculture*, 1949. U.S. Government Printing Office, 1949.

PLANTS

CHRISTENSEN, CLYDE, *The Molds and Man*. University of Minnesota Press, 1951.

COBB, BOUGHTON, *A Field Guide to the Ferns*. Houghton Mifflin, 1956.

MATHEWS, F. SCHUYLER, and NORMAN TAYLOR, EDITOR, *Field Book of American Wild Flowers*. Putnam, 1955.

MOLDENKE, HAROLD N., *American Wildflowers*. Van Nostrand, 1949.

SMITH, ALEXANDER H., *The Mushroom Hunter's Field Guide*. University of Michigan Press, 1963.

WILSON, CARL L., and WALTER E. LOOMIS, *Botany*. Holt, Rinehart and Winston, 1962.

WILDLIFE

ALLEN, DURWARD L., *Our Wildlife Legacy*. Funk & Wagnalls, 1962.

BOURLIÈRE, FRANÇOIS, *The Natural History of Mammals*. Knopf, 1964.

CAHALANE, VICTOR H., *Mammals of North America*. Macmillan, 1947.

LUTZ, FRANK E., *Field Book of Insects*. Putnam, 1935.

MORGAN, ANN H., *Field Book of Animals in Winter*. Putnam, 1939.

MURIE, OLAUS J., *A Field Guide to Animal Tracks*. Houghton Mifflin, 1954.

OLIVER, JAMES A., *The Natural History of North American Amphibians and Reptiles*. Van Nostrand, 1955.

PALMER, RALPH S., *The Mammal Guide*. Doubleday, 1954.

PETERSON, ROGER TORY, *A Field Guide to the Birds*. Houghton Mifflin, 1947.

PETERSON, ROGER TORY, *A Field Guide to Western Birds*. Houghton Mifflin, 1961.

WELTY, JOEL CARL, *The Life of Birds*. Saunders, 1962.

ECOLOGY

ALLEE, W. C. and others, *Principles of Animal Ecology*. Saunders, 1949.

BATES, MARSTON, *The Forest and the Sea*. Random House, 1960.

BENTON, ALLEN H., and WILLIAM E. WERNER, JR., *Principles of Field Biology and Ecology*. McGraw-Hill, 1958.

BUCHSBAUM, RALPH, and MILDRED BUCHSBAUM, *Basic Ecology*. Boxwood Press, 1957.

ELTON, CHARLES, *The Ecology of Animals*. Wiley, 1950.

FARB, PETER, and THE EDITORS OF LIFE, *Ecology*. Time, Inc., 1963.

KENDEIGH, S. CHARLES, *Animal Ecology*. Prentice-Hall, 1961.

MILNE, LORUS, and MARGERY MILNE, *The Balance of Nature*. Knopf, 1960.

NEAL, ERNEST, *Woodland Ecology*. Harvard University Press, 1960.

ODUM, EUGENE P., and HOWARD T. ODUM, *Fundamentals of Ecology*. Saunders, 1959.

OOSTING, HENRY J., *The Study of Plant Communities*. Freeman, 1956.

SHELFORD, VICTOR E., *The Ecology of North America*. University of Illinois Press, 1963.

SPURR, STEPHEN, *Forest Ecology*. Ronald, 1964.

STORER, JOHN H., *The Web of Life*. Devin-Adair, 1960.

THOMSON, BETTY FLANDERS, *The Changing Face of New England*. Macmillan, 1958.

WATTS, MAY THEILGAARD, *Reading the Landscape*. Macmillan, 1957.

SOIL

FARB, PETER, *Living Earth*. Harper, 1959.

RUSSELL, E. JOHN, *The World of the Soil*. Collins, 1959.

U.S. DEPARTMENT OF AGRICULTURE, *Soils and Men, The Yearbook of Agriculture, 1938*. U.S. Government Printing Office, 1938.

THE FOREST TRAVELER

BROOKS, PAUL, *Roadless Area*. Knopf, 1964.

BUTCHER, DEVEREUX, *Exploring Our National Parks and Monuments*. Houghton Mifflin, 1960.

CARRIGHAR, SALLY, *One Day on Beetle Rock*. Knopf, 1944.

DOUGLAS, WILLIAM O., *My Wilderness: East to Katahdin*. Doubleday, 1961.

DOUGLAS, WILLIAM O., *My Wilderness: The Pacific West*. Doubleday, 1960.

EIFERT, VIRGINIA, *Land of the Snowshoe Hare*. Dodd, Mead, 1960.

FROME, MICHAEL, *Whose Woods These Are, the Story of the National Forests*. Doubleday, 1962.

OLSON, SIGURD F., *The Singing Wilderness*. Knopf, 1956.

TEALE, EDWIN WAY, *North with the Spring (The American Seasons)*. Dodd, Mead, 1951.

TEALE, EDWIN WAY, EDITOR, *The Wilderness World of John Muir*. Houghton Mifflin, 1954.

THOREAU, HENRY DAVID, *The Maine Woods*. Various editions.

TILDEN, FREEMAN, *The National Parks, What They Mean to You and Me*. Knopf, 1951.

WILEY, FARIDA A., EDITOR, *John Burroughs' America*. Devin-Adair, 1951.

228

Index

231